Eugen Lamers

# Contributions to Simulation Speed-Up

# VIEWEG+TEUBNER RESEARCH

## Advanced Studies Mobile Research Center Bremen

Herausgeber | Editors:

Prof. Dr. Otthein Herzog
Prof. Dr. Carmelita Görg
Prof. Dr.-Ing. Bernd Scholz-Reiter

Das Mobile Research Center Bremen (MRC) erforscht, entwickelt und erprobt in enger Zusammenarbeit mit der Wirtschaft mobile Informatik-, Informations- und Kommunikationstechnologien. Als Forschungs- und Transferinstitut des Landes Bremen vernetzt und koordiniert das MRC hochschulübergreifend eine Vielzahl von Arbeitsgruppen, die sich mit der Entwicklung und Anwendung mobiler Lösungen beschäftigen. Die Reihe „Advanced Studies“ präsentiert ausgewählte hervorragende Arbeitsergebnisse aus der Forschungstätigkeit der Mitglieder des MRC.

In close collaboration with the industry, the Mobile Research Center Bremen (MRC) investigates, develops and tests mobile computing, information and communication technologies. This research association from the state of Bremen links together and coordinates a multiplicity of research teams from different universities and institutions, which are concerned with the development and application of mobile solutions. The series “Advanced Studies“ presents a selection of outstanding results of MRC’s research projects.

Eugen Lamers

# Contributions to Simulation Speed-Up

## Rare Event Simulation and Short-Term Dynamic Simulation for Mobile Network Planning

VIEWEG+TEUBNER RESEARCH

Bibliographic information published by the Deutsche Nationalbibliothek
The Deutsche Nationalbibliothek lists this publication in the Deutsche Nationalbibliografie; detailed bibliographic data are available in the Internet at http://dnb.d-nb.de.

Dissertation Universität Bremen, 2007

Gedruckt mit freundlicher Unterstützung des
MRC Mobile Research Center der Universität Bremen

Printed with friendly support of
MRC Mobile Research Center, Universität Bremen

1st Edition 2008

All rights reserved
© Vieweg+Teubner | GWV Fachverlage GmbH, Wiesbaden 2008

Readers: Ute Wrasmann | Anita Wilke

Vieweg+Teubner is part of the specialist publishing group Springer Science+Business Media.
www.viewegteubner.de

No part of this publication may be reproduced, stored in a retrieval system or transmitted, in any form or by any means, electronic, mechanical, photocopying, recording, or otherwise, without the prior written permission of the copyright holder.

Registered and/or industrial names, trade names, trade descriptions etc. cited in this publication are part of the law for trade-mark protection and may not be used free in any form or by any means even if this is not specifically marked.

Cover design: KünkelLopka Medienentwicklung, Heidelberg
Printed on acid-free paper
Printed in Germany

ISBN 978-3-8348-0524-9

# Preface

When starting the decision process about the elective courses at the University of Technology in Aachen, the stochastic simulation did not become a favourite candidate in the first place. At the beginning of the main studies, my curiosity somehow made me visit the course and take a look at it. Fascinated, I concentrated on the topic in the following years up to my diploma thesis at the Communication Networks department. At the end of my studies, Prof. Dr. Carmelita Görg offered me to continue with a Ph.D. thesis at the University of Bremen. I willingly accepted, although not having extensively contemplated this before.

I would like to thank Prof. Dr. Carmelita Görg who not only provoked my interest in the stochastic simulation and especially the simulation speed-up. She also managed to intensify the existing excitement and to care about the necessary motivation. The time in her team in Bremen that she is leading in such a family-like way will ever retain a high significance in my life. It has a high value for me not only from scientific perspective, but also personally.

For the review of my thesis I have to thank Andreas Könsgen in particular, and Prof. Dr. Karl-Dirk Kammeyer as the second examiner. During difficult phases of my work I have received a lot of individual support from many persons, of which I would like to mention, apart from all my colleagues and Prof. Dr. Görg, especially Dr. Andreas Timm-Giel. Eventually, achieving this satisfying conclusion of my thesis would have been much tougher if not virtually impossible for me without my wife Nicole, her strength and her heart.

Eugen Lamers

# Vorwort

Als ich anfing, mir im Studium an der RWTH Aachen Gedanken über meine Wahlfächer zu machen, fiel die Stochastische Simulation zunächst einmal aus dem engeren Kandidatenkreis. Als das Hauptdiplom begann, brachte mich meine Neugier dann aber doch irgendwie dazu, der Vorlesung einen Besuch abzustatten. Fasziniert habe ich mich in den darauf folgenden Jahren auf das Thema konzentriert, bis zur Diplomarbeit am Lehrstuhl für Kommunikationsnetze. Am Ende meines Studiums machte Frau Prof. Dr. Carmelita Görg mir das Angebot, die Arbeit an diesem Thema im Rahmen einer Promotion an der Universität Bremen fortzusetzen. Dies nahm ich gerne an, obwohl ich diese Möglichkeit bis dahin nicht näher in Betracht gezogen hatte.

Bedanken möchte ich mich bei Frau Prof. Dr. Carmelita Görg, die nicht nur mein Interesse am Thema der stochastischen Simulation und der Simulationsbeschleunigung im Besonderen geweckt hat. Sie hat es auch verstanden, die vorhandene Begeisterung zu stärken und für die nötige Motivation zu sorgen. Die Zeit in ihrem Team in Bremen, das sie auf so angenehm familiäre Art führt, wird immer einen besonderen Stellenwert in meinem Leben behalten. Sie ist nicht nur aus wissenschaftlicher Sicht, sondern auch für mich persönlich sehr wertvoll.

Für das Review dieser Arbeit möchte ich mich speziell bei Andreas Könsgen sowie für das Zweitgutachten bei Prof. Dr. Karl-Dirk Kammeyer bedanken. In schwierigen Phasen der Arbeit habe ich viel persönliche Unterstützung bekommen, wobei ich hier neben allen meinen Kollegen und Frau Prof. Dr. Görg insbesondere Dr. Andreas Timm-Giel erwähnen möchte. Letztendlich wäre ein so erfreulicher Abschluss der Promotion für mich sehr viel schwerer erreichbar gewesen ohne meine Frau Nicole, ihre Kraft und ihr großes Verständnis.

Eugen Lamers

# Abstract

Starting with an explanation of the general idea of simulation speed-up, first an overview is given of the different simulation speed-up techniques which are examined and compared in this thesis and also investigated with respect to the possibility to combine them in different ways.

Next, the aspect of the state space coverage is discussed and the effect of the local coefficient of correlation is shown. Its influence on the convergence speed of the simulation is demonstrated together with means to reduce this influence significantly, resulting in a considerable speed-up potential.

An introduction is given to the RESTART algorithm to speed-up simulations with focus on rare events. The optimisation of some aspects is discussed, and a technique is shown to be efficient which extends the applicability of RESTART to cases with the only possible thresholds being far away from the optimum.

It is demonstrated how to combine the RESTART algorithm with a parallelisation approach. Different possible approaches are compared and discussed, as well as different possible network architectures and hardware topologies. The chosen approach is explained, and it is shown that there is an actual speed-up on a network of workstations with a high-speed network connection hardware.

A new simulation concept, the short-term dynamic (STD) simulation, is presented. The concept is explained, and it is proven analytically on the basis of a reference queueing model to provide considerable speed-up by reducing the impact of the local correlation. Conceptual ideas are discussed about combining the STD simulation concept with the other speed-up approaches considered in this thesis.

Further, the application of the STD simulation concept to network planning of UMTS is presented. In example simulations of these complex models, the results of the analytic investigation of the concept and the simulations with the reference model are confirmed. The speed-up of the method is shown also for these models, although it is more difficult in such cases to obtain accurate data for the comparison of the simulation performance which can be achieved by the different methods and simulation set-ups.

With the simulation concepts addressed in this thesis, considerable simulation speed-up can be achieved. The application of the presented techniques to other areas is a promising subject for future research.

# Kurzfassung

Beginnend mit einer Erklärung der allgemeinen Idee der Simulationsbeschleunigung wird zu Anfang eine Übersicht über die verschiedenen Simulationsbeschleunigungstechniken gegeben. Diese werden untersucht und verglichen, auch in Bezug auf ihre Eignung zu verschiedenen Möglichkeiten der Kombination.

Als nächstes wird der Aspekt der Zustandsraumabdeckung diskutiert, und der Effekt der lokalen Korrelation wird demonstriert. Deren Einfluss auf die Konvergenzgeschwindigkeit der Simulation wird aufgezeigt sowie Mittel, diesen Einfluss zu reduzieren. Dies führt zu einem beachtlichen Beschleunigungspotential.

Eine Einführung in den RESTART-Algorithmus zur Beschleunigung von Simulationen seltener Ereignisse wird gegeben, sowie die Optimierung einiger Aspekte erörtert. Weiterhin wird eine Technik vorgestellt und für effizient befunden, die RESTART um die Anwendbarkeit auf Fälle erweitert, deren mögliche Schwellwerte weit vom Optimum entfernt liegen.

Es wird vorgestellt, wie der RESTART Algorithmus mit der Methode der Parallelisierung kombiniert werden kann. Sowohl verschiedene Ansätze werden vorgestellt und verglichen, als auch verschiedene mögliche Netzarchitekturen und Hardwaretopologien. Es wird gezeigt, dass der gewählte Ansatz einen tatsächlichen Gewinn erzeugt auf einem Network of Workstations mit schnellem Verbindungsnetz.

Ein neues Simulationskonzept wird präsentiert, das Short-term Dynamic Simulation Konzept (STD). Aus der Basis eines Referenz-Wartenetzmodells wird analytisch bewiesen, dass eine beachtliche Beschleunigung erreicht wird durch die Reduzierung der Auswirkungen der lokalen Korrelation. Weiterhin werden Konzepte zur Kombination mit anderen Simulationsbeschleunigungsansätzen erörtert.

Die Anwendung des STD Konzepts auf Netzplanung für UMTS wird vorgestellt. In Beispielsimulationen dieser komplexeren Modelle werden die Ergebnisse der analytischen Untersuchung und der Simulationen am Referenzmodell bestätigt. Eine Simulationsbeschleunigung ist auch für diese komplexen Modelle erkennbar, obwohl es dort schwieriger ist, verlässliche Werte für Vergleiche der Simulationsgeschwindigkeit zu bekommen, die für die verschiedenen Methoden und Parametrierungen erreicht werden.

Mit den Simulationskonzepten, die hier behandelt werden, können beachtliche Simulationsbeschleunigungen erzielt werden. Die Anwendung der vorgestellten Techniken auf andere Gebiete ist eine viel versprechende Richtung für zukünftige Forschungsarbeiten.

# Contents

# List of Abbreviations

| | |
|---|---|
| **AFAP** | As-Fast-As-Possible (simulation) |
| **ALG** | Average Load Grid |
| **API** | Application Programming Interface |
| **ATM** | Asynchronous Transfer Mode |
| **BHCA** | Busy Hours Call Attempts |
| **CAC** | Call Admission Control |
| **CC** | Congestion Control |
| **CCDF** | Complementary Cumulative Distribution Function |
| **CDF** | Cumulative Distribution Function |
| **CDMA** | Code Division Multiple Access |
| **CLP** | Cell Loss Probability |
| **CNCL** | Communication Networks Class Library |
| **CPICH** | Common Pilot Channel |
| **CRC** | Cyclic Redundancy Check |
| **CRLAR** | Combined Radio Link Addition and Removal |
| **CS** | Circuit Switched |
| **CTMC** | Continuous Time Markov Chain |
| **DES** | Discrete Event Simulation |
| **DIS** | Distributed Interactive Simulation |
| **DL** | Downlink |
| **DoD** | Department of Defence |
| **DTMC** | Discrete Time Markov Chain |
| **FDD** | Frequency Division Duplexing |
| **FDMA** | Frequency Division Multiple Access |
| **GSM** | Global System for Mobile Communications |
| **HLA** | High Level Architecture |
| **IF** | Importance Function |
| **LBS** | Location Based Services |
| **LRE** | Limited Relative Error |
| **ME** | Mobile Equipment |
| **MIMD** | Multiple Instructions, Multiple Data |
| **MMS** | Multimedia Message Service |
| **MoRaNET** | **Mo**bile **Ra**dio **N**etwork **E**valuation **T**oolkit |
| **MPI** | Message Passing Interface |
| **MPP** | Massively Parallel Processor |
| **MuSICS** | **Mu**lti **S**tep **I**mportance splitting **C**lass library **S**ystem |
| **MS** | Mobile Station |
| **NOW** | Network Of Workstations |
| **OE** | Operational Environment |
| **PS** | Partitioned Service (chapter 3) |
| **PS** | Packet Switched (chapters 5 and 6) |
| **QoS** | Quality of Service |
| **RES** | Rare Event Simulation |

**RESTART** REpetitive Simulation Trials After Reaching Threshold

**RLA** Radio Link Addition

**RLR** Radio Link Removal

**RNC** Radio Network Controller

**RND** Random Number Distribution (generator of random number according to a given probability distribution)

**RNG** Random Number Generator (the base generator)

**RRC** Radio Resource Control

**RRM** Radio Resource Management

**RTI** Runtime Infrastructure

**RV** Random Variable

**SCI** Scalable Coherent Interface

**SHO** Soft-Handover

**SIMD** Single Instruction, Multiple Data

**SIR** Signal to Interference Ratio

**SMP** Symmetric Multiprocessor

**SSHO** Softer-Handover

**STD** Short-Term Dynamic

**TDD** Time Division Duplexing

**TDMA** Time Division Multiple Access

**TR** Threshold Refinement

**UE** User Equipment

**UL** Uplink

**UML** Unified Modelling Language

**UMTS** Universal Mobile Telecommunications System

**USIM** UMTS Subscriber Identity Module

**UTRAN** UMTS Terrestrial Radio Access Network

**WCDMA** Wideband Code Division Multiple Access

**WWW** World Wide Web

# List of Symbols

| Symbol | Area | See Page | Meaning |
|---|---|---|---|
| $\Delta I_i$ | RESTART | 36 | distance between thresholds: $\Delta I_i = I_i - I_{i-1}$ |
| $\Delta_{\mathrm{m}}$ | STD convergence | 128 | interval around the (a-priori known) actual mean value for evaluation of STD convergence |
| $\eta$ | queueing systems | 61 | utilization (load): $\lambda/\mu$ |
| $\lambda$ | queueing systems | 61 | arrival rate |
| $\mu$ | queueing systems | 61 | service rate |
| $\rho(x)$ | stochastics, LRE | 68 | local coefficient of correlation at position $x$ |
| $\sigma^2$ | stochastics | 28 | variance |
| $a_i$ | LRE | 17 | transition count from states $< i$ to states $\geq i$ |
| $B$ | RESTART | 20 | rare event of interest |
| $c_{\mathrm{c}}$ | STD simulation | 132 | index of the current event |
| $c_i$ | LRE | 17 | transition count from states $\geq i$ to states $< i$ |
| $c_{\mathrm{l}}$ | STD simulation | 132 | index of the event after which the last recalculation has taken place |
| $d_B$ | RESTART/LRE | 24 | total relative error limit (for rare event $B$) |
| $d_i$ | distr. RESTART | 46 | total relative error limit for step $i$ |
| $d_G(x)$ | LRE | 70 | relative error at position $x$ |
| $d_S$ | RESTART/LRE | 23 | relative error limit per step |
| $\bar{d}_G(x)$ | STD simulation | 72 | expected relative error at position $x$ |
| $\mathbf{e}$ | STD simulation | 132 | event with the index $i$ |
| $g$ | RESTART | 33 | granularity in threshold refinement: number of pieces into which packets are divided |
| $G_x$ | stochastics | 67 | cumulative state probability of all states *right* of $x$ ($1 - G_x$: all states *left* of $x$ including $x$) |
| $\tilde{G}_x$ | stochastics | 70 | obtained cumulative relative frequency of all states *right* of $x$ |
| $h$ | STD analysis | 78 | group number in group correlation analysis |
| $h_i$ | LRE | 17 | absolute frequency of state $i$ |
| $H_i$ | RESTART | 23 | CCDF at threshold $I_i$ with $H_i = G(I_i)$ |
| $H_i^*$ | RESTART | 24 | optimum value for $H_i$ |
| $I$ | RESTART | 20 | threshold |
| $I_{\mathrm{total}}$ | UMTS | 105 | total received interference at base station |
| $k_i$ | RESTART/LRE | 27 | number of traces starting from threshold $I_i$ |
| $K_i$ | RESTART/LRE | 27 | RV of number of traces starting from threshold $I_i$ |
| $m$ | RESTART | 22 | index of highest threshold $I_m = B$ |

| Symbol | Area | See Page | Meaning |
|---|---|---|---|
| $M$ | distr. RESTART | 46 | number of slave processes |
| $N_{\mathrm{g}}$ | STD analysis | 84 | number of groups in group correlation analysis |
| $n_i$ | RESTART/LRE | 27 | number of trials needed in step $i$ |
| $\tilde{n}_i$ | RESTART/LRE | 27 | estimated number of trials needed in step $i$ |
| $N_i$ | RESTART/LRE | 27 | random variable of number of obtained trials in step $i$ |
| $n_{i,k}$ | distr. RESTART | 46 | number of trials obtained in step $i$ by slave $k$ |
| $n_{\mathrm{s}}$ | simulation | 70 | total number of samples obtained during simulation |
| $n_{\mathrm{s,g}}$ | STD analysis | 83 | fix number of samples obtained within a group |
| $\overline{n}_{\mathrm{s,w}}$ | STD simulation | 74 | mean number of samples (evaluated states) in STD window |
| $N_{\mathrm{w}}$ | STD simulation | 74 | number of STD windows in simulation |
| $\mathbf{p}$ | Markov chains | 78 | general transition probability matrix, one-step |
| $\overline{\mathbf{p}}(n_{\mathrm{s,g}})$ | STD analysis | 84 | $n_{\mathrm{s,g}}$-step mean transition probability matrix |
| $\mathbf{P}(s\|x_0)$ | STD analysis | 85 | conditional probability matrix for the state sum $s$ depending on starting state $x_0$ |
| $\mathbf{P}(x_n\|s)$ | STD analysis | 86 | conditional probability matrix for the starting state $x_n$ of the next group depending on a state sum $s$ of the current group |
| $\mathbf{P}(x_n\|s,x_0)$ | STD analysis | 86 | conditional probability matrix for the starting state $x_n$ of the next group depending on a state sum $s$ and the starting state $x_0$ of the current group |
| $p_0(x)$, $p_1(x)$ | Markov chains | 66 | transition probabilities of the equivalent 2-node Markov chain (calculation of local correlation $\rho(x)$) |
| $p_{\mathrm{ex}}$ | STD simulation | 73 | probability of extra transition (STD window transition) |
| $\mathbf{p}_g$ | STD analysis | 87 | transition probability matrix for state sums |
| $p_{ij}$ | stochastic | 64 | transition probability from state $i$ to state $j$, one-step |
| $p_{ij}(n)$ | stochastic | 78 | $n$-step transition probability from state $i$ to state $j$ |
| $P_{\mathrm{max,DL}}$ | UMTS | 105 | maximum transmission power in downlink |
| $P_{\mathrm{N}}$ | UMTS | 105 | background and receiver noise at base station |
| $p_{\mathrm{s}}$ | STD analysis | 64 | death probability (next departure before next service) |
| $P_{\mathrm{total,DL}}$ | UMTS | 105 | total downlink transmission in downlink |
| $r_i$ | LRE | 17 | cumulative frequency of the states $< i$ |
| $R_i$ | RESTART | 22 | splitting factor, i. e., number of retrials per transition state at threshold $I_i$ |
| $R_i^*$ | RESTART | 24 | optimum value for $R_i$ |
| $s$ | STD analysis | 85 | state sum of a group in group correlation analysis |
| $s$ | state coverage | 8 | step size, level of state space quantisation |
| $\mathbf{s}$ | STD analysis | 85 | vector of relative frequencies of state sums in group correlation analysis |

| Symbol | Area | See Page | Meaning |
|---|---|---|---|
| $S$ | STD analysis | 85 | random variable of the state sums in group correlation analysis |
| $S_i$ | RESTART/LRE | 27 | random variable of the number of trials in a trace of step $i$ |
| $t_{\mathrm{i}}$ | STD convergence | 128 | duration of initial phase |
| $t_{\mathrm{rel}}$ | STD convergence | 128 | relative duration the simulated mean value has to stay uninteruptedly within interval $\Delta_{\mathrm{m}}$ around actual mean value |
| $t_{\mathrm{s}}$ | simulation | 61 | total simulation time |
| $T_{\mathrm{w}}$ | STD simulation | 74 | size of STD window (duration) |
| $\upsilon_x$ | stochastic | 70 | cumulative frequency of all states *right* of $x$ |
| $\mathbf{v}$ | STD analysis | 78 | steady state probability vector |
| $\mathbf{v}_0$ | STD analysis | 84 | relative frquency vector of starting states of groups |
| $\overline{\mathbf{v}}^{(h)}$ | STD analysis | 83 | mean probability vector for the states in group $h$ |
| $\mathbf{v}_k^{(h)}$ | STD analysis | 78 | probability vector for the $k^{\mathrm{th}}$ state in group $h$ |
| $\mathbf{v}_0^{(h)}$ | STD analysis | 78 | starting probability vector for group $h$ |
| $\upsilon_i$ | LRE | 17 | cumulative frequency of the states $\geq i$ |
| $w(\mathbf{e}_i)$ | STD simulation | 132 | weight of the event $\mathbf{e}_i$ |
| $X$ | stochastic | 20 | general random variable, e. g. occupancy |
| $\mathbf{x}(n_{\mathrm{s,g}})$ | STD analysis | 84 | vector of mean state values depending on the starting state |
| $\overline{x}^{(h)}$ | STD analysis | 85 | mean state of group $h$ |
| $\overline{x}$ | STD analysis | 85 | total mean state |
| $x_0$ | STD analysis | 85 | starting state of a group |
| $x_n$ | STD analysis | 86 | starting state of the next group |
| $X_t$ | stochastics | 28 | stochastic process |

# List of Figures

# List of Tables

# Chapter 1

# Introduction

Communication networks of today must meet high demands regarding the Quality of Service (QoS). Simulative performance evaluation of communication networks is an important means for the design and configuration of such networks. Other methods of performance evaluation are the measurement of an existing real system and the analytic evaluation of a model describing the target system. Measurement is not possible in the design stage of a system, and analytic methods are usually restricted to very simple models of the target system neglecting some or many possibly important aspects.

In simulation, the complexity of the model and the level of details can be defined flexibly, depending on the aspects the investigation is focused on. Simulations of more complex models, however, are slower than simple models, i. e., they require more run time per simulation time unit.

But the question is now, what is meant by slow or fast simulation. It has to be clarified what is meant by speed-up, and how it can be defined. Assume that results of a reference simulation are available and results from another simulation which has been performed with a certain speed-up technique. The question remains how these techniques can be compared, and some comparison parameter is needed.

This comparison parameter has to be taken from some sorted value space to map the state of the simulation progress to. The progress state the reference simulation has reached when finished, the reference progress state, is mapped to a certain reference value of that value space. The simulation progress state of the other simulation to which the reference simulation is compared also needs to be mapped to that same value space. In fact, for each progress state during the simulation a value is needed. These values are intended to reflect the statistical accuracy, and on the basis of these values, the simulations can be compared. If the test simulation reaches the reference progress state in less time than the reference simulation, the test simulation is rated as the faster simulation, or the simulation technique is the faster one.

Applying a speed-up technique to a simulation now means, a reference statistical accuracy can be reached in shorter time, or a higher statistical accuracy can be reached within the reference time. The simulation user knows the constraints for conducting simulations. Either the user has a maximum of available time for simulation, or a minimum statistical accuracy is required to manifest a certain statement. In any case, a speed-up technique can either save time or increase statistical accuracy.

The criterion for the statistical accuracy used in this thesis is the relative error for a given range of the value space of the evaluated random variable. The so-called local correlation coefficient makes a significant contribution to this relative error. Chapter 2 explains the effect of the correlation on the state space coverage and the intrinsic simulation speed-up potential resulting from the systematic reduction of the correlation. Further, a short introduction on parallelisation is given and the evaluation algorithm Limited Relative Error (LRE) is explained which is of fundamental importance for the investigations of all other chapters.

Special speed-up techniques for the simulation of rare events are indispensable. Otherwise, reliable simulation results would not be obtainable within reasonable time. A technique called importance splitting and especially a variant of it called RESTART is introduced in chapter 3. It treats regions of the value space differently in order to reach the relative error limit for each region in a comparable time.

Another speed-up technique is parallel processing. It uses a number of processing units to work on different parts of the simulation task concurrently. Basic principles of parallelisation are given at the end of chapter 2, and a combination with the RESTART technique is discussed in chapter 4. It is different from the parallelisation of straightforward simulations, since the RESTART algorithm itself is divided into autonomous parts which can run concurrently.

The short-term dynamic (STD) simulation concept is introduced in chapter 5, and it is compared to the concepts of static and dynamic simulation. Its capability to reduce the local correlation and in turn to speed-up the simulation, is investigated analytically on the basis of a simple queueing model.

The way the STD simulation concept is applied to the network planning for the Universal Mobile Telecommunications System (UMTS), is shown in chapter 6. The concept of a simulation toolkit designed for this simulation concept is demonstrated. Simulations are presented showing the potential and limitations of the STD concept.

Finally, in chapter 7, conclusions are drawn about the different concepts presented in this thesis, and possible further combinations are evaluated.

# Chapter 2

# Statistical Accuracy

In simulation, the goal is either to achieve a most accurate estimate of the target characteristic of the system within a certain time, or to reach an estimate with a certain accuracy as fast as possible. Improving a simulation method to speed-up the simulation, or finding a new method, increases the statistical accuracy. It is up to the user whether he profits from the achievement by getting the same results as before, only faster, or by getting better results within the same computation time. Thus, whenever something is denoted as simulation speed-up, instead it could be called increasing the statistical accuracy. The term *variance reduction* is frequently used for techniques increasing the statistical accuracy.

## 2.1 Discrete event simulation

In discrete event simulation (DES), a model describes some real world system. During the simulation, at certain points in time $t_i$ the current system state is collected. Dependent on the investigated $n$-dimensional random variable (RV) $X$, the corresponding $n$ parameters of the system state – possibly out of more than $n$ – are obtained meaning that the resulting $n$-dimensional value $x_i$ is saved for statistical evaluation. *Save* in this context does not necessarily mean that the value is stored in memory for later usage, but can also mean immediate processing, depending on the way the statistical evaluation is conducted. Obtaining a single value $x_i$ is called an *observation*.

The distribution function describing the random variable $X$ is estimated by the distribution of the relative frequency of the observations, resulting from all the $x_i$. Within the $n$-dimensional state space $\mathbb{Z}^n$ of the simulated model, corresponding to the discrete $n$-dimensional RV $X$, every possible state $x$ has its steady state probability $P_X(X = x) > 0$ in a steady state simulation. For a continuous RV $X$, every possible $n$-dimensional area $x$ of the of the continuous $n$-dimensional state space $\mathbb{R}^n$ has its steady state probability $P_X(X \in x) > 0$, and $P_X(X = x) = 0$ for a continuous state space.

The accuracy of an estimate is reflected by the variance of the sample provided that the target characteristic is the sample mean.

It is more sophisticated if the target is the distribution function of the sample, since an error measure has to be introduced which provides information about the estimate accuracy at every possible position of the value space.

There are two major techniques for rare event simulation. One is the importance sampling which changes the probability distribution of the simulated model to have events of interest occur more frequently. The other is importance splitting, especially RESTART which is explained in chapter 3. In this method, states that are closer to the states forming the (rare) event of interest are stored in order to restart the simulation from such states several times. The fact that conditional probabilities are obtained by this method has to be incorporated into the final calculation of the absolute probability of the event of interest.

## 2.2 State space coverage

To achieve a good estimate of the distribution function of $X$, the absolute frequency of every possible state should be sufficient (see section 2.4). If the frequency is zero, the state would even be considered as an impossible state. With an infinite or practically infinite state space, resulting from either a large number of dimensions or a large value space within the dimensions, it may not be possible to fulfil this requirement. The same applies for a continuous value space and, furthermore, for the case of a rather small number of possible states but with some states having extremely low probability. As a consequence, stochastic simulation techniques are required to study such systems.

For evaluation purposes, the value space of a one-dimensional RV can either be discrete or, for continuous RVs, the value space is divided into intervals of appropriate size, depending on the problem, memory considerations, the requested accuracy, etc. This means, the determination of a distribution function of a continuous one-dimensional RV also must be made discrete. Thus, the value spaces are all considered to be discrete in the following. Continuous value spaces are divided into regions for which the probability can be estimated, and are transformed into discrete value spaces by making a single value of this interval represent the interval, e. g., a boundary value or the mean value. Value spaces which are already discrete can be further quantised by combining intervals. In the more general case of $n$-dimensional RVs with $n > 1$, value regions have to be considered.

From the evaluation perspective, a region is considered to be a single value from the discretised value space representing this region. The value space can be assumed to be covered well as soon as all discrete values reach a predefined error limit, see section 2.4.

### 2.2.1 Effect of correlation

First, the influence of the correlation on the coverage will be discussed and visualised with the help of example simulations. The correlation is a measure for the degree of similarity of two random variables. The term *correlation* used in this thesis always means the first order auto-correlation which means the similarity is considered between two consecutive values of a stochastic process. Values for the coefficient of correlation are in the range of $[-1; +1]$. Positive values mean a high similarity for consecutive values while negative values mean dissimilarity. A coefficient of correlation of zero indicates completely unrelated values.

In most cases in which the correlation is addressed, a global coefficient is considered. In this thesis, the local coefficient of correlation $\rho(x)$ is considered instead. It reflects the fact that the degree of similarity between consecutive values also depends on the value (the location) $x$. As an example, in a stochastic process there can be a value $x_i$ which is followed always by values very close to $x_i$, and the values which follow the value of $x_j$ are rather unrelated to $x_j$. This means, $\rho(x_i) \approx 1$, and $\rho(x_j) \approx 0$.

#### 2.2.1.1 Uniformly distributed state space

Figure 2.1 shows for three different simulation approaches the observations that have been obtained during example simulation runs. In all these simulations, the stationary distribution function of the states is uniform over the chosen and displayed value region of $[0,180] \times [50,170]$ of the continuous value space which is dimensionless for simplicity. To make the difference between these approaches visible, the simulated sample size $n_s$ was chosen to be quite small with $n_s = 2000$.

The points in the figure represent positions of elements, and the lines represent the movements of the elements. This has been chosen to visualise the simulation principles. A very simple mobility model has been used which allows an element to start at a random position and at every movement to turn backwards, sidewards, or keep the direction. The travelled distance is fixed. It has been taken care that the stationary spacial distribution of the element positions is the same for all simulation examples.

The first simulation approach, figure 2.1 a), is a Monte-Carlo simulation, also called *static* simulation in the following. In a sequence of observations over the simulation time, all observations are independent of the previous ones. In spite of the small sample size, the values (points) are spread over the whole area almost uniformly.

The second approach, figure 2.1 b), is called *short-term dynamic* simulation (STD). Here, the simulated sequence of observations consists of a sequence of sequences (groups). Such a sequence of observations (a group) is called an STD window, and the length of the sequence is defined by the simulation user, either in time units or number of observations, or even randomly following some probability distribution.

Within such an STD window, the observations are not independent, leading to correlated sequences. The observations within an STD window are independent of those of other STD windows. If STD windows are considered as units, an STD simulation can be considered as a Monte Carlo simulation of these units since subsequent units are independent.

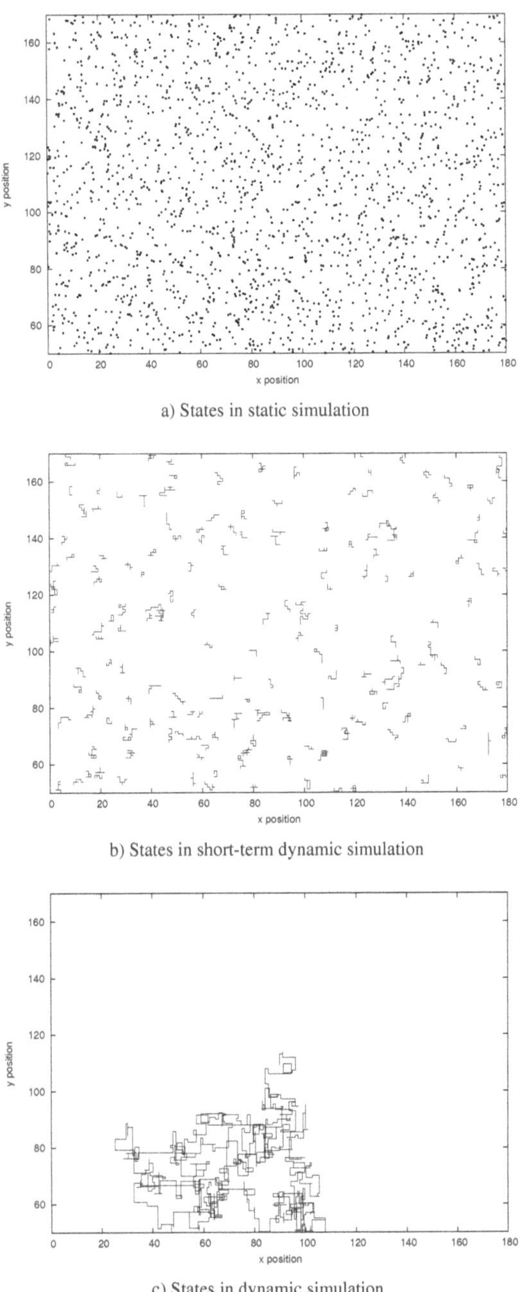


a) States in static simulation

b) States in short-term dynamic simulation

c) States in dynamic simulation

**Figure 2.1:** Uniformly distributed states

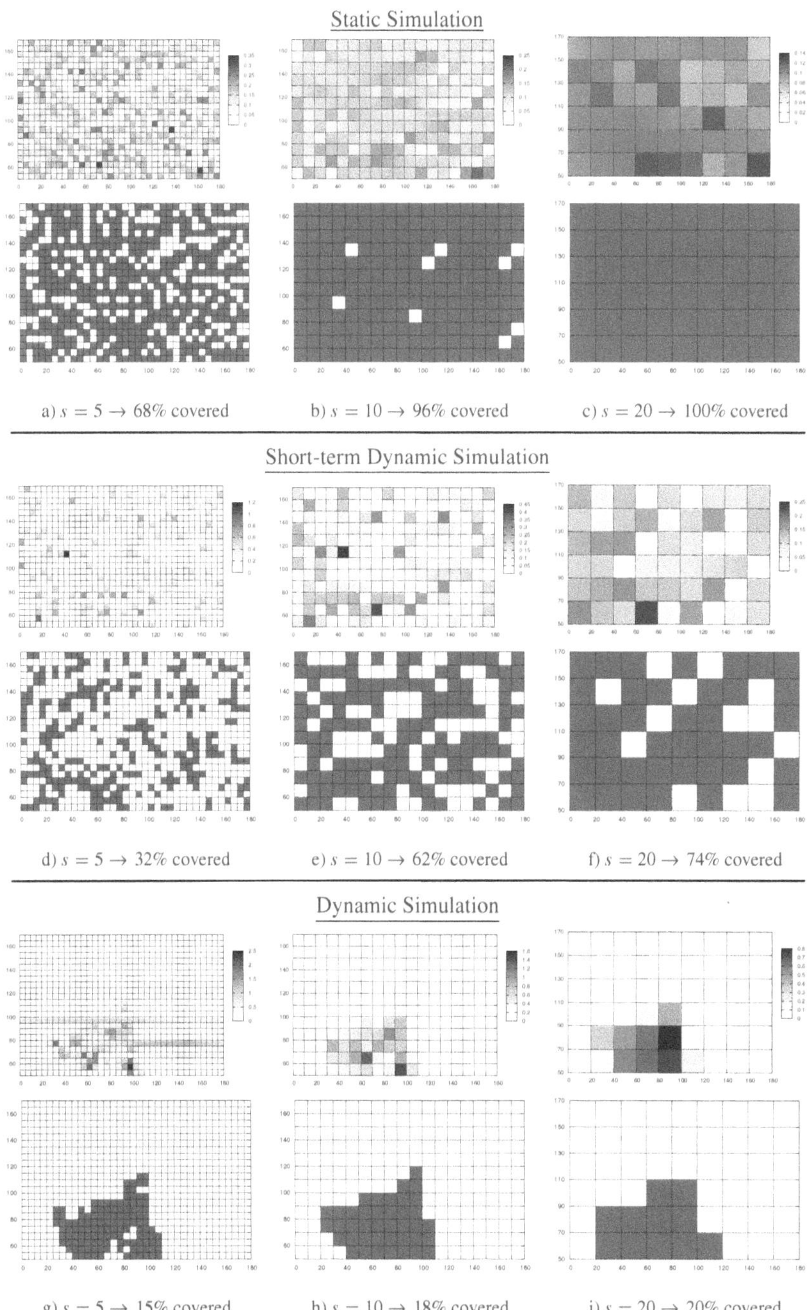


**Figure 2.2:** Covered areas for uniformly distributed states

This approach will be further discussed and analysed in chapter 5. Figure 2.1 b) shows these groups of subsequent values represented by short lines. The groups themselves are spread over the whole area, also quite uniformly. Considering the points, the distribution over the area is less uniform than for the static simulation example.

The third approach, figure 2.1 c), is called *dynamic* simulation. As in the short-term dynamic simulation, the sequence is correlated, but there is only a single long sequence instead of a number of STD windows.

Figure 2.2 shows how differently these approaches cover the value space of a simulated random variable. It shows combining parts of the value space to square areas. Different levels of quantisation are shown, indicated by a step size $s$. Here, $s = 5$ means squares of the size $5 \times 5$ units of the value space are combined, equivalently for $s = 10$ and $s = 20$. The upper part of each of the three blocks of the figure shows the mean number of observations per unit square within the combined square areas. Darker grey mean higher values. The lower parts show for a threshold value of 0.08, which squares represent sufficiently covered areas. This applies to a square if the mean number of observations per unit square ($1 \times 1$) within that square area is larger than the threshold. The very small threshold is due to the small sample size and is in this example about 86 % of the expectation value for a unit square.

In case of a static simulation as in figure 2.1 a), in the long run all possible values will be observed equally often. For small levels of quantisation, large sample sizes (long simulations) are needed to keep the frequency of observations of all possible values similar. The corresponding parts of figure 2.2, which are 2.2 a), 2.2 b) and 2.2 c), show this effect. For $s = 10$ and $s = 20$, (almost) all areas are covered, for $s = 5$ only less than 70 % of the squares are covered. Obviously, for $s = 5$ the sample size is far too small. The homogeneity of the coverage, however, is high.

For the short-term dynamic simulation approach, the corresponding parts figures 2.2 d), 2.2 e) and 2.2 f) show that the coverage is lower than for the static simulation. The sample size for this approach is too small for all quantisations shown, even for $s = 20$, but an increasing tendency can be clearly seen. And also here, the homogeneity of the coverage is quite high.

In the dynamic simulation approach, the simulation keeps for a while within a small area around the starting value due to the highly dependent states. For the example simulation in figure 2.1 c), figures 2.2 g), 2.2 h) and 2.2 i) show that for the short simulation time the simulation run covers only a small part of the value space. Even larger quantisations do not help very much since most parts of the value space have not been visited at all.

**Dynamic Model**

The dependent sequences generated for the graphs of the dynamic and the short-term dynamic simulation represent moving objects in a 2-dimensional area. The directions to which the objects move in the next step are determined with respect to predefined turning probabilities. These probabilities are chosen in such a way that the steady state probabilities for all positions of the area are the same as for the static simulation. This representation has been chosen to visualise correlated sequences for a 2-dimensional random variable.

The starting point of a sequence if drawn from a uniform distribution the same way as every point in the static simulation. From there, the process kind of walks over the value space by deciding on the direction and distance of the next step. The underlying mobility model must be parametrised in such a way that in the long term the relative frequency of all possible states matches the steady state distribution function of the starting points. For long-term comparability, the boundary conditions must also be taken into considerations. For these examples, this is neglected because very short runs are executed to make the correlation effect visible. The mobility model follows the same principle as the mobility model introduced in section 6.2.3, and here it is used with simplifications.

#### 2.2.1.2 Tandem queue

In figure 2.4, a simple tandem-queue model as shown in figure 2.3 is simulated. The service processes are Markovian, as well as the arrival process. The buffer sizes are infinite, the loads are $\eta_0 = 0.95$ for the first queue and $\eta_1 = 0.9$ for the second queue.

**Figure 2.3:** Simple tandem queue

On the x-axis, the occupancy of the first system (queue and server), and on the y-axis, the occupancy of the second system is shown. The state is evaluated whenever an arrival occurs from the outside, i. e., to the first system. For the short-term dynamic and the dynamic simulation, figures 2.4 b) and 2.4 c), subsequent observations are connected with lines. The density of these lines gives an impression of the frequencies of the corresponding areas of the state space. The frequency of a single value cannot be reflected by this way of representation, which is why only a small sample size is simulated to distinguish between visited and unvisited states. In the case of the static simulation, figure 2.4 a), the problem occurs again more seriously. It is resolved by having the type and the size of the points reflect the observed frequency of the state. In the legend of figure 2.4 a), the corresponding relative frequency for each point type and size is shown. In contrast to the uniformly distributed states in the previous section, the probability of the states with small or even zero occupation of both systems is higher than for large occupancies. The theoretical mean values are

$$E[X] = 19, \qquad E[Y] = 9 \tag{2.1}$$

with $X$ and $Y$ being the RVs for the occupancies of the first and the second system.

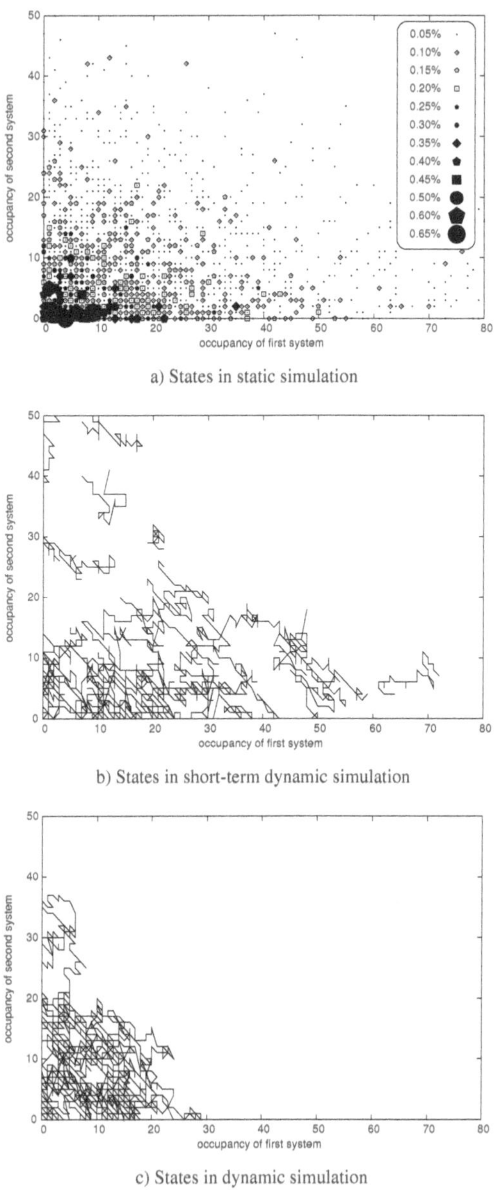


a) States in static simulation

b) States in short-term dynamic simulation

c) States in dynamic simulation

**Figure 2.4:** Queue occupancy states of tandem model

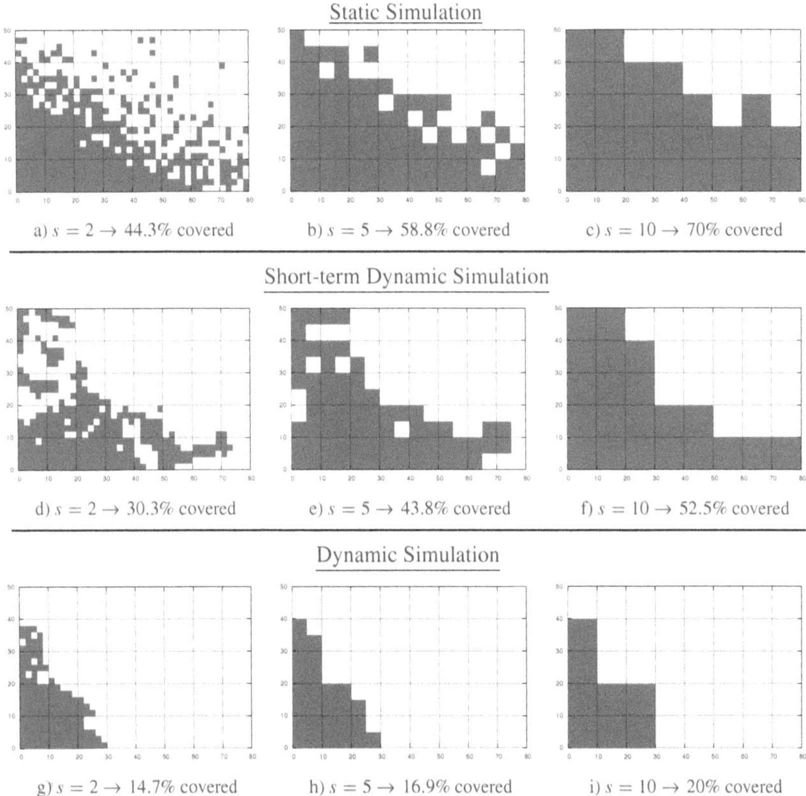


tandem queue loads $\eta_0 = 0.95$ and $\eta_1 = 0.9$, 20 time units per STD window,
$\approx$ 2000 values simulated, threshold (coverage) 0.05 per value

**Figure 2.5:** Covered areas for tandem queue simulation

In figure 2.5, the covered areas of the considered state space are shown for the different simulation approaches and different levels of quantisation. Approximately $n_s \approx 2000$ values are simulated, and the frequency counters are scaled to $n_s$ to achieve comparable results.

The results are in line with the results of figure 2.2. The coverage ratio increases with the quantisation level and from dynamic over short-term dynamic to static simulation. The setup of the example simulations is chosen within the limits beyond which the coverage ratio will not change further. This applies to the quantisation level and to the threshold.

#### 2.2.1.3 Interrupting the correlation

From the simulations it can be seen that the simulation approach with the long correlated random sequences, the dynamic simulation, has a lower coverage ratio for the same sample size than the simulation approaches with the shorter correlated sequences, i. e., the short-term dynamic simulation and especially the the static simulation with the completely uncorrelated random sequence. This is the reason for the simulation speed-up here.

Approaches to interrupt the correlation by scrambling the sample of a dynamic simulation would obviously not lead to any success. In the evaluation procedure, the LRE method (see section 2.4) would indeed report lower correlation for the sample and, as a consequence, a lower relative error for the values which have been observed. But this is only because the evaluation procedure is betrayed by providing it a different system. And further, the coverage ratio does not increase, irrespective of the quantisation level.

In dynamic simulations, usually very well known states are chosen as the starting state, mostly an empty system. Provided that it is a valid state and the simulation produces a large enough sample, the choice of this starting state will not have a relevant influence on the error of the simulation, even if the steady state probability of this state is quite low.

For a short-term dynamic simulation, starting with an arbitrary state or always with the same state, e. g., the state that would be chosen for the dynamic simulation, would introduce an unknown bias. As in the static simulation, the distribution functions of the *input* parameters must be known. Input parameters can be, e. g., number, position and status of objects, while the *output* parameters for which the simulation is conducted, are compound values like sums, mutual influence between objects, etc. Further examples are discussed together with simulations of mobile radio networks in chapter 5 and especially chapter 6.

### 2.2.2 Rare events

Figure 2.6 shows the probability density function of a continuous value space. For simplicity, the value spaces are assumed to be dimensionless in the following. Figure 2.6 a) shows the 2-dimensional graph of the function which is $z = s^2 \cdot \exp(-\alpha(x + y))$. Here, $s$ is the step size as introduced in section 2.2.1, and $\alpha$ is a parameter of the negative exponential function determining the decay rate. The highest value for the area $x > 0$, $y > 0$ is $z = \alpha^2$ at $x = y = 0$,

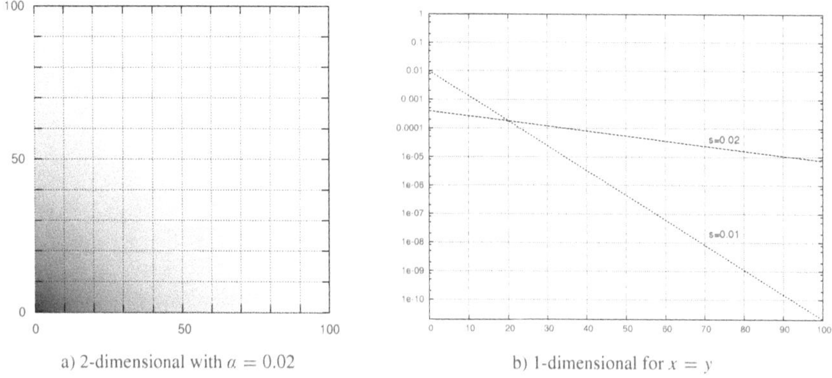


a) 2-dimensional with $\alpha = 0.02$ b) 1-dimensional for $x = y$

**Figure 2.6:** Probability density function: $z = f(x,y) = \alpha^2 \cdot e^{(-\alpha(x+y))}$

and the values for $z$ are represented by shades of grey, with black representing $z = \alpha^2$. Figure 2.6 b) shows a one-dimensional extract of the two-dimensional density function for $x = y$ and for two different values of $\alpha$.

In a simulation in which a random variable is evaluated which is distributed as in the graph, in the long term the frequencies of observations in all regions will correspond to the distribution function. The events which lead to observations in the upper right part of the graph are rare, the lower left region will be observed rather often. To reach a certain error level, a minimum number of observations for each considered discrete value must have been made. This minimum value increases with decreasing requested error limit.

Figure 2.7 shows three different quantisations. The left diagrams have square regions of the size $5 \times 5$, the middle have $10 \times 10$ squares, and the bottom diagrams have $20 \times 20$ squares. In each square, the values of the corresponding square of the left diagram are summed by integration. If the sum exceeds a threshold, the square is painted grey, otherwise it is kept white. A grey square represents a sufficiently evaluated discrete value for a given error level.

In this example, the threshold has been chosen to be 15 times the highest value at $x = y = 0$. The upper parts show the function for $\alpha = 0.02$. For the different quantisation levels $s$, 5, 10, and 20, the considered value space has a coverage ratio of 3.8 %, 45 %, and 96 %. The gain in coverage is achieved with a loss of precision by a factor of 4 between the quantisation levels. With $\alpha = 0.02$, there is indeed a gain, and with $\alpha = 0.1$, there is no gain. The higher regions of the value space have much lower probability than for $\alpha = 0.02$ resulting in only the lowest left square being covered, no matter what quantisation level is chosen.

In chapter 3, a method is introduced by which it is possible to make regions with very low probability being visited and covered well within reasonable amounts of time. This is achieved by storing system states every-time the stochastic process crosses a predefined threshold in the direction towards the low probability region (the rare event). If the stochastic process crosses

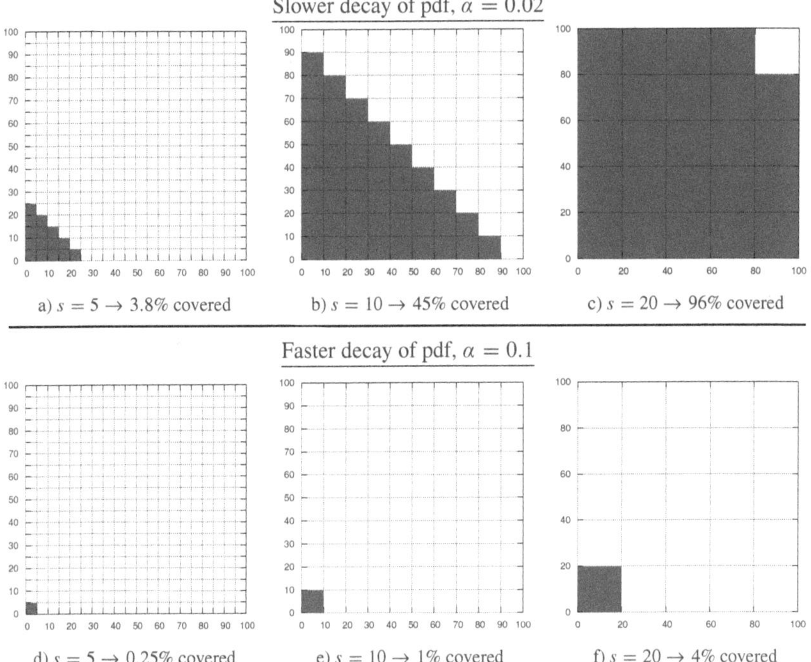


**Figure 2.7:** Levels of value space quantisation

back over this threshold, one of the stored states is restored from which the simulation proceeds. The conditional probabilities which this approach is based on have to be incorporated in the calculation of the total probability of the rare event.

## 2.3 Parallelisation

Parallelisation does not address the relation between the simulation accuracy and the sample size, but the division of the simulation work into parts which can be done concurrently on a number of computers, or, more generally, processes.

First, parallelisation can mean a single sequence of instructions (a program) for multiple parallel sequences of data. This is called *SIMD* (Single Instruction, Multiple Data). From a programmers' point of view, this is still sequential computing. Since this kind of parallelisation is not appropriate for typical stochastic simulation problems, the speed-up is limited.

With the other type of parallelisation, autonomous processors with dedicated instruction streams are working on dedicated data streams. This is called *MIMD* (Multiple Instructions, Multiple

Data), and it is further referred to as distributed computing. Only this type is considered further in the following and in chapter 4.

Although the processes have their own program and data, in distributed simulation they all work on the same large problem. This means that somehow all parts have to interact. How much and what kind of interaction is necessary, depends on the problem. Further, the parts that have to interact must be enabled to do so, which means, that the structure of the interactions (the set of programs) must fit to the connection structure. This is the communication platform and is usually some kind of network or shared memory. The need for interaction leads to the need to synchronise the processes. The efficiency of the simulation suffers from synchronisation loss resulting from situations in which one of the processes has to pause until an interaction with some other process can take place, see [Fuj00].

Three different systematics classify the parallelisation of simulations. They are introduced briefly in the following.

### 2.3.1 Functional division

The functional division considers the different functions from the simulators' view. A simulator needs to generate (pseudo-)random number sequences, to evaluate and possibly to post-process the obtained results and to visualise them. From each of these elements more than one can exist. But there is only one control program which represents the actual simulation while all other processes represent auxiliary functions. This limits the possible speed-up, but the synchronisation loss is low. Special processors can be used for the different tasks.

### 2.3.2 Load division

The load division is a set of complete and independent simulation runs with different parametrisations and/or different seeds of their random generator(s) which are executed in parallel on processors which generally do not even need to interact. In the case of only different seeds, the results of the simulation runs all representing the same model with the same parametrisation can be combined afterwards. In contrast to functional division, universal processors are needed here which can perform all kinds of tasks.

### 2.3.3 Model division

The model division assigns components of the model to different logical processes which can run on separate processors. An important issue for this approach is how the components are logically connected. This means what change in the state of a certain component is relevant for the process of a certain other component. These relevant change events have to be communicated and form the interaction structure of the model. The model time of the interacting components has to be synchronised. For efficiency reasons, the components should be allowed to work autonomously as long as possible, otherwise the loss due to components waiting for other components to send information is high.

Optimistic approaches assume that components can work autonomously for quite a long time. They assume that the probability is low for the need to react on relevant events of other components. This leads to lower communication overhead, but the system must be able to rollback the state of the components to a point in time at which a relevant event occurred. This can be time consuming if it happens frequently for a large number of components. Pessimistic approaches only go forward to a certain point in model time for a component if it is definitely known that all possible relevant events from other components will not be scheduled at an earlier point in model time. This increases the frequency of components waiting for information from other components. Furthermore, if this interaction structure is not carefully implemented, it can easily lead to deadlocks, especially for more complex models.

### 2.3.4 Efficiency

The complexity of the simulation with respect to the required interaction between the parts of the work determines how efficiently the divided work can be completed. In the load division, the required communication will be low leading to high efficiency and speed-up. In the functional division, the communication will also be quite low. The efficiency depends on how many random numbers are needed per time unit, how many (single) results are produced, and how fast these informations can be communicated. In the model division, the communication effort depends strongly on the model and the choice of either an optimistic or a pessimistic approach.

Further discussion on parallelisation approaches is done in chapter 4 with application on the rare event simulation speed-up technique RESTART.

## 2.4 Statistical evaluation with LRE

The algorithm called *Limited Relative Error* (LRE) [GS90], [SG96] is used for statistical evaluation. It not only provides an estimation for the probability of a special event of interest, but also for the complete complementary cumulative distribution function (CCDF) and measures for the relative error and the local correlation of each possible value of the value space. The LRE measures the correlation within the sample sequence and incorporates this into the control of the simulation run time, since higher correlations lead to lower state space coverage ratios (section 2.2) and require larger samples sizes.

Since the above mentioned estimations are accessible during the simulation, the LRE can also be used for simulation control. This includes on the one hand simply the question whether the total number of observations (the size of the obtained sample) is large enough to fulfil the error requirements given by the user and to finish the simulation. On the other hand, in the RESTART/LRE algorithm (chapter 3), the LRE decides, whether a part of the simulation – a step – is sufficiently evaluated, and further serves as a decision making process for finding the next threshold automatically with the help of the intermediate distribution function calculated by the LRE.

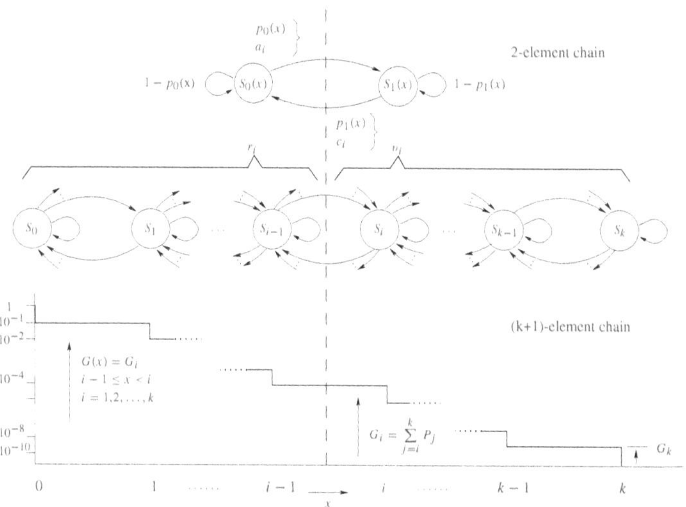


**Figure 2.8:** Markov chain for LRE [SG96]

Figure 2.8 shows a Markov chain with finite number of elements representing a one-dimensional state space with a finite number of states $(k+1)$. The state space needs to be one-dimensional because the states need to be ordered (every state needs two neighbours except from the boundary states). For each state, two counters are managed. The first is the absolute state frequency $h_i$ which counts how often the state $i$ has occurred. The cumulative frequencies $v_i$ (figure 2.8) are calculated as follows: $v_i = \sum_{j=i}^{k} h_j \quad \forall i = 0, 1, \ldots, k$ with $v_0 = n$, and $n$ is the total size of the simulated sample. The second counter is the transition frequency $c_i$ which counts how often transitions from states $j$ with index $j \geq i$ to states with index $j < i$ have occurred.

To gain reliable results, the so-called large sample conditions of the LRE in equations (2.2), (2.3) and (2.4) have to be fulfilled. $r_i, v_i$ are the cumulative state frequencies of the states $0 \ldots i-1$ resp. $i \ldots k$ with $n = r_i + v_i$, and they are needed to determine the state probabilities $S_{0/1}(x)$ of the 2-element Markov chain, see figure 2.8. $a_i, c_i$ are needed to determine the transition probabilities $p_{0/1}(x)$ of the 2-element Markov chain.

An example is given in section B explaining the meaning of the large sample conditions in more detail, which are:

$$n > 10^3 \tag{2.2}$$

$$r_i > 10^2 \quad \wedge \quad \upsilon_i > 10^2 \tag{2.3}$$

$$a_i > 10 \quad \wedge \quad c_i > 10 \quad \wedge \quad r_i - a_i > 10 \quad \wedge \quad \upsilon_i - c_i > 10 \tag{2.4}$$

The other equations for the measured CCDF $\tilde{G}(x)$, which are the simulated mean value $\tilde{\alpha}$ regarding the state of the Markov chain, the measured local coefficient of correlation $\tilde{\rho}(x)$, the corresponding correlation factor $\widetilde{cf}(x)$, and the relative error $d_G(x) = \sigma_G(x)/\tilde{G}(x)$, are as follows with $i - 1 \leq x < i$ and $i = 1 \ldots k$:

$$\tilde{G}(x) = \tilde{G}_i = \upsilon_i/n \tag{2.5}$$

$$\tilde{\alpha} = \frac{1}{n}\sum_{i=1}^{k} \upsilon_i \tag{2.6}$$

$$\tilde{\rho}(x) = \tilde{\rho}_i = 1 - \frac{c_i/\upsilon_i}{1 - \upsilon_i/n} \tag{2.7}$$

$$\widetilde{cf}(x) = \widetilde{cf}_i = \frac{1 + \tilde{\rho}_i}{1 - \tilde{\rho}_i} \tag{2.8}$$

$$d_G(x) = d_i = \sigma_G(x)/\tilde{G}(x) = \sqrt{\frac{1 - \upsilon_i/n}{\upsilon_i} \cdot \widetilde{cf}_i} \tag{2.9}$$

The local correlation $\rho(x)$, as introduced at the beginning of section 2.2.1, indicates the degree of similarity of the value $x$ with the subsequent value in the stochastic process. The correlation factor $\widetilde{cf}(x)$ is included in the relative error and reflects the contribution of the local correlation to the relative error.

# Chapter 3

# Simulation Speed-up with RESTART

In every simulation domain there is the goal to achieve the most accurate results within the shortest possible run-time, see chapter 2. The basic idea and the necessity of simulation speed-up has been discussed. In the domain of rare event simulation (RES), simulation speed-up techniques are of particular importance, since otherwise no useful results would be obtainable within an appropriate amount of time. A comparison of some essential simulation speed-up techniques is given in [Hee95].

In this chapter, the principles of the technique RESTART are introduced. Many investigations have been conducted with the simulation toolkit MuSICS (**Mu**lti **S**tep **I**mportance splitting **C**lass library **S**ystem), initially developed by Oliver J. Fuß. Several extensions have been developed during this work to enable the investigations presented in this chapter. More details are found in [GLA99], [LG00] and [LG04]. Investigations with distributed RESTART are shown in chapter 4.

## 3.1 RESTART fundamentals

In 1991, J. and M. Villén-Altamirano have introduced this technique [VAVA91b] as an enhanced variant of importance splitting which has been suggested by A. J. Bayes in 1970 [Bay70]. The simulated random variable is considered in such a way that its value space is divided into importance regions which can be handled separately. In [GS96], [Gör97] and [GF98], the possibilities of the combination of RESTART and LRE are shown.

### 3.1.1 Motivation

To investigate QoS in high-speed networks, e. g., cell loss probabilities (CLP) in ATM networks in the range of $10^{-9}$, special simulation techniques are required which are able to evaluate the statistical properties of such rare events with a sufficient accuracy.

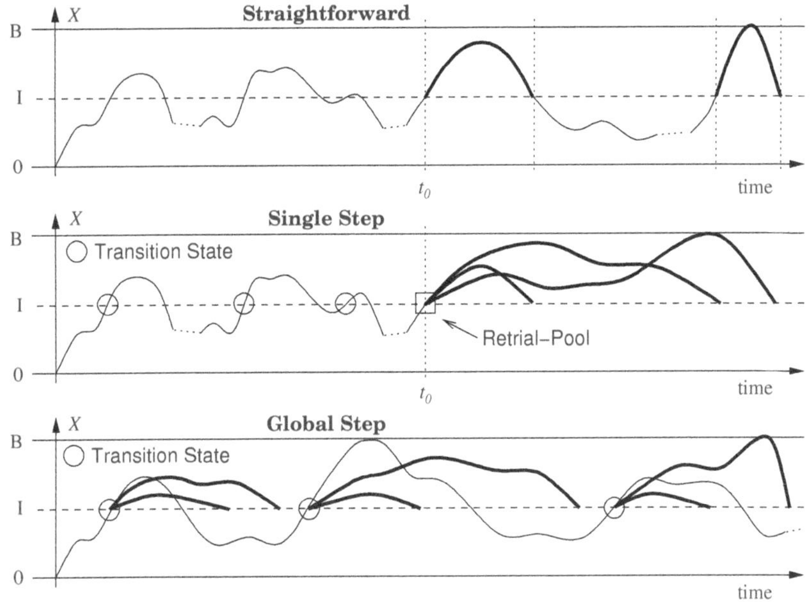


**Figure 3.1:** RESTART approaches

Figure 3.1 shows examples of simulation traces for a straightforward simulation and for RESTART simulations. A trace shows the observations of a stochastic process plotted over time representing the random variable $X$. The simulation executes this stochastic process by identifying the current value of the random variable at certain points in time. These time instants can be, e. g., the arrivals of new packets at the evaluated network node.

Thresholds divide the value space of $X$ is into importance regions. In this example, there are two importance regions and a single threshold $I$ represented by the horizontal dashed line. The importance region in the interval $[0, I]$ has a higher probability than the interval starting at $I$. The rare event of interest is $B$, and all values $x \geq B$ form the rare event set.

A lower probability of the upper region leads to a lower sample size in the upper region than in the lower region. A minimum sample size for each region is required to fulfil given error conditions.

In the straightforward simulation in figure 3.1, the first vertical dashed line represents the time $t_0$ at which the evaluation of the lower part of the random variable in the interval $[0, I]$ is evaluated sufficiently with respect to its error condition. The dots in the developing of the values indicate a large amount of model time during which the random variable remains below the threshold $I$.

After this point of time, the evaluation does not need further values for $x < I$, only the values $X \geq I$ are needed. Values of the lower region would increase the accuracy of that region, however, this is no longer needed. Thus, the time spent in simulation producing values for the lower region after the time $t_0$ is wasted. Consequently, in RESTART the simulation of the lower importance region after fulfilling its error condition at $t_0$ is skipped in favour of the next higher region. The intervals between the thresholds and the limits are called *steps* and match the importance regions.

### 3.1.2 Requirements

The idea of RESTART is based on the conditional probability $P_{B|I}$ of $B$ for $X > I$:

$$\begin{aligned} P_{B|I} &= \mathrm{P}\{x \geq B \mid X > I\} \\ P_{B|I} &\geq P_B \end{aligned} \tag{3.1}$$

Generally, on its path from the starting state at $x = 0$ to the rare event of interest $B$, the stochastic process has to visit the intermediate states, including the threshold $I$. This leads to the relation $P_0 > P_I > P_B$. Further, this is why it can be expected to reach the rare event faster from an intermediate state like the threshold $I$ than from the starting state or other states below $I$. To take advantage of this, $B$ must be reachable from $I$ with $X > I$, meaning the process does not need to go below $I$ to reach $B$ afterwards.

E. g., if the system occupancy is evaluated, an empty system has a high probability, an occupancy of 20 has a comparably low probability, and an occupancy of 10 has a medium probability. Starting with the state with the high probability, at some time before the state with the low probability can be reached, the state with the medium probability must have been reached before. In this example, it is $B = 20$ and $I = 10$.

Further discussion on the topic, whether there can be jumps over one or more thresholds and the impact on the efficiency, can be found in [Gar00] and [VA98].

### 3.1.3 Procedure

To enable the favouring of importance regions, the simulation must stop at the time when a transition from the current step to the next lower step is performed. Steps represent the important regions and are called steps since they are ordered. At that transition, the simulation has to restart with a trace which tends to go up again in order to remain in the current importance region and to approach values closer to the rare event of interest. The required upward tendency can be provoked by storing the system state of the simulation every time a threshold $I$ is reached by an upward transition from the step below to the step above the threshold $I$. The circles in figure 3.1 indicate such situations. The system states which are stored are called *transition states*. Picking up (restoring) a previously stored transition state and starting the simulation again from that state is called a *retrial*.

Storing promising states with a tendency to the rare event and restarting from them in the next higher step, forms the actual splitting. The name comes from the fact, that each transition state is reused several times which splits the trace leading to the state into several traces. This can be best seen at the global-step part of figure 3.1 where this is visible at every transition state. The number of times the transition state is used, defines the splitting factor $R_i$ at threshold $I_i$. For the single threshold in the figure it is $R_0 = R = 3$.

The difference between the *single-step* and the *global-step* approaches is as follows: In the single-step method, only a single-step is simulated at a time. In the first phase, the first step is simulated, and the transition states for the second step are collected. After finishing the first step, the second step is simulated with retrials starting from selected transitions states of the retrial pool. In a multiple step case, transition states for the third step are collected during the simulation of the second step. In the global-step all steps are simulated concurrently. The order of collecting a state and starting retrials is changed and reflects a recursive procedure. Both approaches can have multiple steps and are implemented in MuSICS.

### 3.1.4 System state

A system state can have many parameters depending on the simulated model. In a simple single server queue, the system state consists of the number of packets or jobs in the queue, the remaining service time of the currently served packet, and the remaining time until the next packet is generated. If the service or the generation process is memory-less, the corresponding remaining time does not need to be stored. For an M/M/1 model, only the system occupancy including queue and server has to be stored, resulting in a one-dimension state space. The state space of a less simple model is usually multi-dimensional.

Attention has to be paid to the (pseudo) random number generator (RNG). It is impossible for RESTART to provide any speed-up if the state of the RNG is also stored together with the system state. The state of the RNG is either an inner state of the generation algorithm or the position in the random number sequence in case of using a stored sequence. Storing this state would lead to exact copies of the simulation traces starting from the same transition state.

### 3.1.5 Threshold setting

Now, the more general multi-step case is considered. It has more than two steps and, by that, more than a single threshold. These thresholds are called $I_i$ with $I_0$ always being the threshold above the first step and $i = 0,..,m$ with $I_m = B$, see figure 3.2. This setting consists of $m + 1$ steps and also $m + 1$ thresholds.

For RESTART with only one single-step, as in figure 3.1, the basic equation is $P_B = P_{B|I} \cdot P_I$. For the multi-step case, it is as follows:

$$P_{I_i} = P_{I_i|I_{i-1}} \cdot P_{I_{i-1}} \tag{3.2}$$

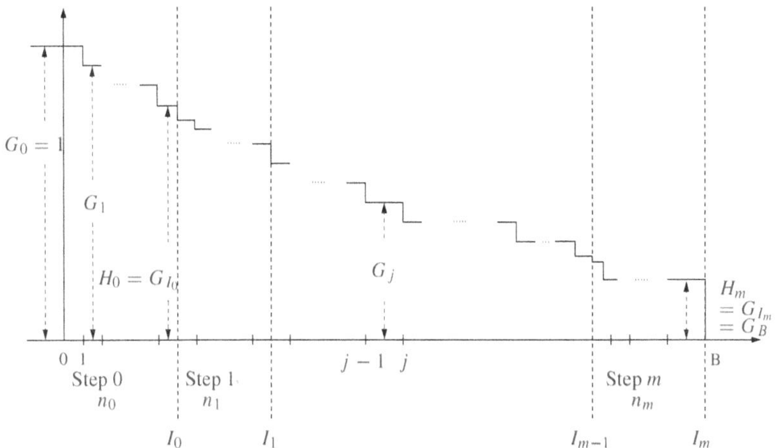


**Figure 3.2:** Splitting the CCDF for multi-step RESTART [Gör97]

The probability of the rare event or, more generally, the rare event set $B$ is $P_B = \mathrm{P}\{X \geq B\}$ and is given as the product of the conditional cumulative probabilities of the thresholds:

$$P_B = G_B = P_{I_0} \cdot \prod_{i=0}^{m-1} P_{I_{i+1}|I_i} \tag{3.3}$$

Figure 3.2 shows on the $x$-axis the value space of the random variable from $x = 0$ to $x = B$. Larger values are not included since the value at $x = B$ represents $\mathrm{P}\{X \geq B\}$, and for cases in which values of $x > B$ are possible, all values $x \geq B$ are considered as the rare event set, and there is no concern for further fragmentation of the area $x \geq B$. The example value space displayed in the figure is discrete leading to the stepped distribution function. It represents the Markov chain of all possibles states of the stochastic process. The Markov chain is the Markov chain of the corresponding LRE, see figure 2.8 in section 2.4. The areas between the thresholds $I_i$ and between 0 and the threshold $I_0$ represent the $m + 1$ steps.

The distance between the levels $H_i = G(I_i)$ of neighbouring thresholds should be set as close to optimum as possible. If no further knowledge about model dependent parameters is taken into account, see [PLG02] and the discussion in section 3.4, the distance should be equal for all steps if the same step error limit $d_S$ is defined leading to the same number of trials per step:

$$\frac{H_{i+1}}{H_i} \approx \frac{H_i}{H_{i-1}} \quad \text{with } H_i = G(I_i) \tag{3.4}$$

A larger number of steps leads to a higher relative error for $B$, provided that the steps are all simulated with the same error limit $d_S$. This is because every factor in the product of equation

(3.3) contributes to the total error. In equation (3.5), the relation between the maximum error $d_B$ for $B$ and the per step error $d_S$ is shown:

$$d_B = \sqrt{\left(d_S^2 + 1\right)^{m+1} - 1} \approx d_s\sqrt{m+1} \tag{3.5}$$

Discussions about the optimum values $R_i^*$ and $H_i^*$, the resulting number of steps and the sample size, depending on correlation and given maximum error, can be found, among others, in [SG94], [GS96], [Gör97], and [PLG02].

## 3.2 Importance function

In the above examples, the parameter system occupancy has been used as the random variable describing the stochastic process. The values of this parameter have been used to indicate the system states, and thus, this parameter controls the RESTART activities of saving and restoring the transition states. The thresholds reside within the value space of the parameter, and the rare event set is described using the dimensions of the parameter.

In general, the state space of the model under investigation only needs to be mapped onto a real-valued parameter, the so-called importance function (IF). This is represented by this parameter in the above examples. The term has been introduced in [VAMMGFC94]. The only requirement for an optimum importance function for a one-dimensional state space is that it is monotone increasing, see [VAMMGFC94] and [Gar00].

### 3.2.1 Multivariate importance functions

In a multi-dimensional state space, which is the case for most not purely theoretical models, however, not every importance function is efficient, see the discussions in [VAVA99]. This makes it necessary to incorporate more than one parameter of the system state into the control of RESTART, i. e., to enable multivariate importance functions in MuSICS.

The evaluation of the IF is performed at discrete times, for which one network node object of the simulated model is responsible. E. g., if the IF is simply the arrival occupancy at node **A**, then this node is responsible for putting the occupancy of its queue into the evaluation process every time a new job has arrived. For the node this is straightforward, since the information about the queue occupancy is available to the node.

If parameters from more than one node need to be collected for the IF, e. g., the sum of the occupancies of node **A** and node **B**, one node still must be responsible for the decision about the discrete times when the values are collected and used for the IF. Such times can be the events of arrivals to that node. Two approaches to collect the parameters from all involved nodes are considered. Let **A** be the responsible node.

1. At every time the responsible node **A** wants to calculate the current value of the IF, it asks all involved nodes explicitly for their contributions, which is a polling mechanism.
2. Every time the contributed values change at the involved nodes except the responsible node (in this case only node **B**), the node provides the updated information to the responsible node **A**. At the time node **A** wants to calculate a new value of the IF, it uses the available information.

The disadvantage of the first approach is that node **A** needs access to all other nodes. It is easier to provide access to the responsible object to all nodes, which favours the second approach. Similar is the application of a global object to which all involved nodes write their updated parameter information and from which the responsible node reads the information when needed.

The second approach, however, can be less efficient if there are frequent updates of the contributed parameters in node **B**, and this frequency is significantly higher than the updates in node **A**.

In the inverse case, where the update frequency in node **A** is significantly higher, the second approach avoids parsing a list of nodes although their values have not changed since the last inquiry. This list is, however, usually very small. An IF with a high number of involved parameters would require very deep knowledge of the modelled system and would have to be investigated as a separate research topic. Thus, the second approach with the global object for the contributed parameter information is in general the better solution, even if in special cases the other approach can be more efficient.

### 3.2.2 Random variable of interest

Generally, the simulation user is not interested in the evaluation of a multivariate random variable which values are available just because the IF represents it for efficiency reasons. Now a network is considered in which the occupancy of some queue is the random variable $X$ while the occupancy of some other queue is the random variable $Y$. For efficiency reasons, in this example, the IF has been chosen to be some function of both $X$ and $Y$, namely $Z = f(X,Y) \geq 0$. The random variable of interest, however, is $X$. Now, with $X > x_R$ as the rare event of interest, $n+1$ thresholds $l_i$ and the law of total probability

$$\sum_i P(B_i) = 1 \Rightarrow P(X) = \sum_i P(X \cap B_i), \tag{3.6}$$

it follows

$$P(X \geq x_R) = \sum_{i=0}^{n+1} P(X \geq x_R \cap Z \in [l_{i-1}, l_i]) \quad \text{with } l_{-1} = 0,\, l_{n+1} = \infty \tag{3.7}$$

In each step $i$, a $P(Z \geq l_i)$ is determined, which is the basic procedure in RESTART/LRE. Additionally, as an element of the sum in equation (3.7), a $P(X \geq x_R \cap Z \in [l_{i-1}, l_i])$ is determined in each step $i$ by calculating the relative frequency of events $X \geq x_R \cap Z \in [I_{i-1}, I_i]$ in step $i$. Since this leads in the first place to the conditional probability $P(X \geq x_R \cap Z \in [I_{i-1}, I_i] | Z \geq I_{i-1})$, it has to be multiplied by $P(Z \geq I_i)$.

The same way as for the RV $Z$, also for $X \geq x \cap Z \in [I_{i-1}, I_i]$ a cumulative distribution function $G(x)$ can be evaluated with the LRE. It has to be taken care about the fact that simulated observations for which $Z \in [I_{i-1}, I_i]$ does not apply, have to be ignored for inclusion into the $X$-LRE, while they are considered for the $Z$-LRE to enable the RESTART/LRE on the basis of $Z$ as the importance function.

### 3.2.3 Merging LREs

Each element of the sum in equation (3.7) represents one RESTART step and, by that, a single LRE evaluation object of the random variable $X$ for the corresponding step. The $I_i$, however, are the thresholds of the random variable $Z$ which is the IF for controlling the RESTART algorithm.

The intervals within the $X$-LREs have to be the same in all levels, to enable merging the complete distribution function for $X$. If only the probability of the rare event $P(X \geq x_R)$ is of interest, this is not necessary, but one interval representing the rare event has to start at the same value, preferably at $x_R$.

## 3.3 States in single-step

In the global-step method, a trace which runs between the thresholds $I_{i-1}$ and $I_i$ and hits the upper threshold $I_i$ is split immediately according to the splitting factor $R_i$. In single-step, the transition states are collected until the current step is finished and the next step is simulated. Some aspects about these state collections are now considered.

### 3.3.1 Required size of collection

A certain number of transition states collected in the current step are needed for the simulation of the next step. The number $R_i$ of new traces that will be started within the next step needs to be estimated to make statements about the required collection size. $R_i$ is a given parameter for step $i+1$ (between thresholds $I_i$ and $I_{i+1}$).

Figure 3.3 shows a part of a simulation currently being in step $i$ between thresholds $I_{i-1}$ and $I_i$. The consecutive traces are plotted over the simulation time which is in the following expressed in trials instead of time units.

The traces have different lengths which is not directly affected by whether the upper threshold $I_i$ is hit or not. This means, the trace is not terminated at the upper threshold as it is done in the global-step method where immediately the new transition state is used to split the successful trace. Thus, the length of the traces is independent of the choice of the upper threshold.

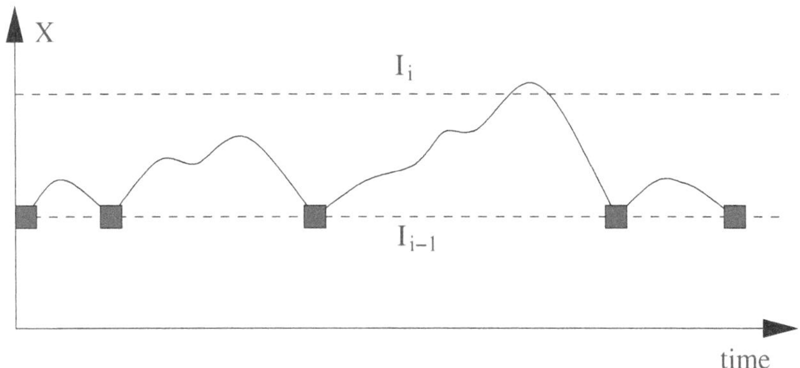


**Figure 3.3:** Traces in single-step

The total number of trials needed in step $i$ is $n_i$. The number of obtained trials in step $i$ as a random variable $N_i$ results from the number of traces $k_i$ starting from threshold $I_i$ and from the number of trials $S_i$ of which these traces consist. $S_i$ is a discrete random variable. The discrete-time stochastic process $S_{i,k}$ is stationary, since the sequence of traces represent a renewal process with the renewal point shown by the squares in figure 3.3. Here, $k$ is the sequence number of the trace. The stationarity leads to a sequence of i. i. d. random variables $S_{i,k} = S_i$. This results in

$$N_i = \sum_{j=0}^{k_i} S_i = k_i S_i \tag{3.8}$$

The question, however, is not the total number of trials given a fixed number of traces, but the random variable $K_i$ which is the number of traces needed to obtain $n_i$ samples in step $i$. The total number of trials in step $i$ can be estimated incorporating the requested error level and further information about the expected correlation. The actual number of trials will not be predetermined, but there are some possibilities to estimate it like short pre-simulations with higher error. Thus, it is now assumed to have an estimate $\tilde{n}_i$. The random variable $K_i$ is now

$$K_i = \tilde{n}_i \frac{1}{S_i} \tag{3.9}$$

In typical applications, the state probabilities decrease with increasing importance function which is a real-valued representation of these states. This leads to a lower probability of longer traces. Further, it can be assumed that the traces are independent if they do not emerge from the same transition state. Long term dependencies due to sharing of the same transition state one or more thresholds further down are not taken into account in this consideration. Since generally no other attributes of the trace length distribution is known, it is assumed to be geometric reflecting the independence, the non-uniformity, and the fact that it has a minimum of 1 without an upper limit.

The expectation of the trace length $S_i$ is $E(S_i) = \bar{S}_i = p_i^{-1}$ with $p_i$ being the success probability of the geometric distribution, and the variance is $\sigma^2_{S_i} = \bar{S}_i^2 - \bar{S}_i$. The probability mass function for $K_i$ is then

$$\begin{aligned} P_K(K_i = k) &= P_S(S_i = s) \quad \text{for } s = \frac{n_i}{k} \\ \Rightarrow P_K(K_i = k) &= P_S(S_i = \frac{n_i}{k}) = p_i(1-p_i)^{\frac{n_i}{k}-1} \end{aligned} \tag{3.10}$$

Under the assumptions made and chosing an optimum $R_i$, the number of traces $K_i$ leads to the number of states $Z_i$ needed in step $i$. It is $Z_i = \frac{1}{R_i} K_i$ with the expectation

$$E(Z_i) = \frac{1}{R_i} \sum_{k=1}^{\infty} p(1-p)^{\frac{n_i}{k}-1} = \frac{1}{R_i} \frac{p}{1-p} \sum_{k=1}^{\infty} (1-p)^{\frac{n_i}{k}} \tag{3.11}$$

The most important parameter $p$ of the distribution of $Z_i$ is still unknown. Furthermore, the type of distribution of $S_i$ can be slightly different from the geometric distribution. This means that even if the mean value of $S_i$ is known, it remains difficult to calculate a good estimate for $\bar{K}_i$ without having a certain knowledge about the stochastic process $X_t$ the sample is obtained from. Pre-simulations of representative models for the investigated class of models, however, can provide reasonable information about the required parameters. Also already gathered experience with similar models can be included into the estimation.

In contrast to $S_i$, an estimate for the needed samples $\tilde{n}_i$ can be concluded from the step which is currently running. From the progress information of the achieved error levels, the number of trials still needed to finish the step can be derived. This is because the number of needed trials is proportional to the reciprocal square root of the maximum error.

### 3.3.2 State usage and limitation

The condition for the successful finish of step $i$ is reaching the specified error level including the fulfilment of the large sample conditions, see section 2.4. At the end of the simulation of step $i$, the resulting total number of trials in that step is $n_i$. The number of states for the next step $i+1$ which have been collected in step $i$ can be lower than expected as discussed in section 3.3.1. It can also be very high which leads to a large number of states that will not be used in the next step or at least to a low usage factor $R_i$ of the states for step $i+1$ at the threshold $l_i$.

From the view of simulation efficiency, the effect is small. The number of operations to restore a state in the next step is independent of the state collection size, at least if the underlying memory structure is not implemented in a sequential-access like manner for random access. Only the number of operations to store a state is higher. Since the prediction of the needed number of states for the next step is not trivial to do, as discussed in section 3.3.1, it is better not to stop storing states in step $i$ and continue until the simulation in step $i$ is finished.

Another aspect is the memory consumption for the collection of the transition states. High memory consumption can, in some cases, reduce the performance of a process, especially if swapping to mass storage devices is involved. If the number of collected states exceeds the necessary number for the next step by orders of magnitude, a limit to the collection can be applied.

Pre-simulations, as discussed at the end of section 3.3.1, help to estimate the required number of states. The fact that the resulting $R_i$ is independent of the specified error level, the pre-simulations can be conducted with a relaxed error level.

If these results show, e. g., an effective resulting $R_{i,\text{res}} = 0.5$ while the predefined one would be the optimum value $R_i = R_i^* = e^2 \approx 7.4$, only less than one of about ten states needs to be stored. This can be done by stopping to store transition states after the estimated number of states has been collected. Acting like this will lead to the situation that only the transition states of the earlier simulation phases of step $i$ are stored which can possibly have impact on the accuracy of the results. A better way is to store only the first state in a group of ten consecutive states.

### 3.3.3 State selection

From the retrial-pool, the collected states can be selected in a linear way or randomly. In the linear way, the order in using the collected states is not changed from the order in which they have been collected. When reaching the end of the pool, it is started over at the beginning forming a circular selection procedure.

First, an optimum size of the state collection is assumed which will result in a good $R_i$. In this case, the linear selection is the preferred one. There are no additional random operations needed, and all states are reused equally making the linear selection kind of fair. More precisely, it means, each state is used either $\lfloor R_i \rfloor$ or $\lceil R_i \rceil$ times. Even in cases of quite a low $R_i$ of about 2 or 3, the selection fairness remains valid, and thus, the preferred selection is the linear one.

On the other hand, if the $R_i$ is very low, e. g., significantly below 2, the size of the state collection is higher than necessary, and a state limitation should be considered, see section 3.3.2. The simulation should be conducted again, or at least the limitation could be applied to planned simulations with similar parameters.

If the state collection size is very high and causes the operating system to do memory swapping, the simulation should be aborted in any case if detected. The efficiency in such cases can easily go far below the efficiency of straightforward non-RESTART simulations.

If system memory is non-critical and a state limitation is not necessary because the efficiency gain is expected to be small, see discussion in section 3.3.2, the random selection is to be preferred. It is because it does not favour the earlier simulation phases from which all reused states have been collected disregarding completely all states that have been collected in later simulation phases.

A badly configured simulation can have very small $R_i$. The only reason can be that the probability to hit the upper threshold from the lower threshold is relatively high. This will lead to a small number of required samples in that step and, consequently, to a small number of required retrials. Optimally, $R_i$ should be as close as possible to the reciprocal value of the conditional probability $P_{I_{i+1}|I_i}$ to reach threshold $I_{i+1}$ for $X > I_i$:

$$R_i \approx P^{-1}_{I_{i+1}|I_i} \tag{3.12}$$

This is because a number $n_i$ of trials in step $i-1$ lead to a number of states $k_i$. If only one retrial from each of these $k_i$ states would be made, there would be collected

$$n_{i+1} = n_i P_{I_{i+1}|I_i} \tag{3.13}$$

trials in step $i$. Here, a Poisson process is assumed if there is no further knowledge about the process. If the thresholds are set optimally, the number of trials needed in the steps are equal: $n_i = n_{i+1}$, and thus, equation (3.12) follows.

The reason for a badly configured simulation can either be that equation (3.12) has not been applied, or that the conditional probabilities $P_{I_{i+1}|I_i}$ have been estimated badly. If all this is not the case and the simulation is well configured, the linear selection should be preferred because of the prevention of additional computation overhead.

### 3.3.4 Combination of single-step and global-step

A combination of the single-step method with the global-step method can use the advantages of both methods. As identified above, the single-step method consumes more memory as the global-step method. The amount of required memory depends on the properties of the simulated system defining the memory consumption for each transition state. But it can increase significantly if the $R_{i,\mathrm{res}}$ is very low leading to a large number of states to be stored. In contrast to this, in the global-step method only a single state per threshold needs to be stored at a time.

In the global-step method, however, the thresholds need to be defined in advance. This can lead to a significantly decreased efficiency if the thresholds are far away from the optimum values. The single-step method is able to collect a certain number of observations in a first phase of a step without collecting transition states. From the distribution function resulting from this phase an optimum threshold can be defined. In a second phase, the transition states at the threshold which is now defined, can be collected. The first phase runs with a higher error limit of the LRE than the second phase. Apart from the error limit, the LRE evaluation is the same. This provides the possibility for the simulation user to achieve a threshold selection close to optimum.

Combining these method means, in a first phase the single-step method finds the optimum thresholds, and in the second phase, the global-step method runs using these thresholds. The single-step simulation can run with a higher error limit, because it is only a pre-simulation phase to determine the optimum parameters of the main simulation running with the global-step method.

## 3.4 Threshold refinement

Not all scenarios allow for the application of the optimum parameters of the RESTART technique. One aspect is the geometric distance of the thresholds on the probability axis. Equation (3.14) shows this as a relation of the levels $H_i$ of the thresholds $I_i$:

$$H_{i|X \geq I_{i-1}} = \frac{H_i}{H_{i-1}} \quad \forall i = 0, \ldots, m \tag{3.14}$$

The theoretical optimum is $H_{i|X \geq I_{i-1}} = e^{-2}$ as long as no unknown model dependent parameters are taken into account, see [PLG02].

If the random variable is discrete, obviously the thresholds $I_i$ cannot always be set to the exact values corresponding to the optimum $H_i^*$. They have to be rounded to the closest value of the value space. Precisely, a threshold can also be put between two possible neighbouring values of the random variable's value space, but not more than one. Otherwise, no transition between these thresholds would be possible, since always a transition would involve all these thresholds together in one step.

In most cases, this will be sufficient. If the distance between two possible values for the thresholds is large, however, at least for a region of the value space, this will cause a low efficiency for the affected region.

As an example, figure 3.4 shows a random variable with a CCDF of $G(x) = e^{-10 \cdot x}$. Possible thresholds for a RESTART simulation of this random variable are the natural numbers 1,2,3,..., indicated by the thick dashed vertical lines. They have a distance of $H_{i|X \geq I_{i-1}} = G(x|X \geq x-1) = e^{-10}$. Thus, the optimum thresholds would be at 0.2,0.4,0.6,..., indicated by the thin dashed vertical lines.

### 3.4.1 Partitioned service

In the event driven simulation applied in MuSICS, a packet generator (source) determines the inter-arrival time values drawn from the RND and sends the next packet to the receiving server node. This packet, however, does not know its size or duration, it is up to the RND of the

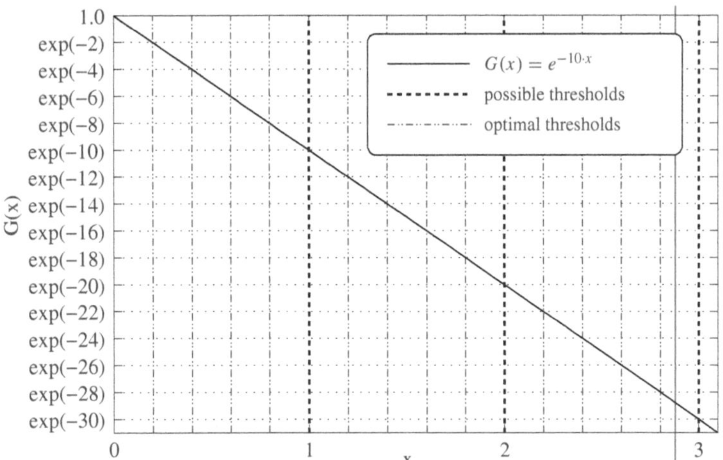


**Figure 3.4:** Threshold refinement

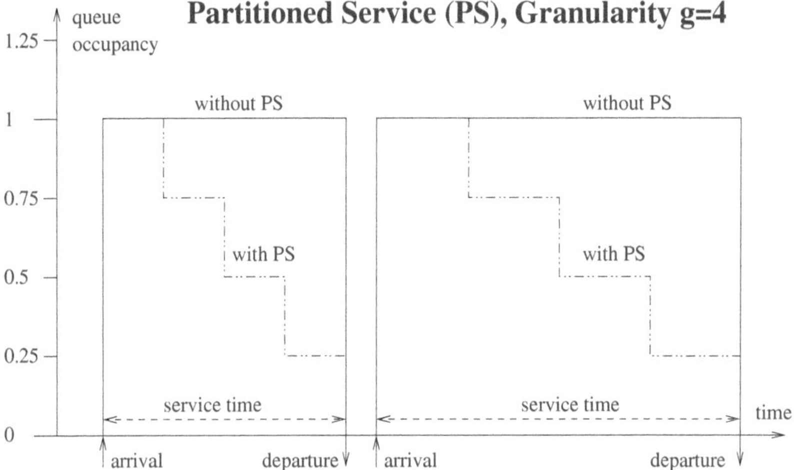


**Figure 3.5:** Partitioned service

receiving station managing the service process by which the decision about the service time is made. The simulation-relevant point of time at which the packet is sent further to the next node resp. removed from the network, is determined by this service time.

As a consequence, the station receiving a packet has to take care about the refinement of the random variable's value space (the queue occupancy), which in reality matches (a subset of) the natural numbers. It starts at zero for an empty system and ends at the queue length plus one for the servers, or the number of servers if it is a multi-server node.

At the event indicating the end of the service time of the currently served packet, the station removes the packet from the system or forwards it to the next receiving node. In case of waiting packets in the queue, the next service period starts at the end of the previous service period, and otherwise it starts at the arrival of a new packet from the source. The duration of the new service period is again determined by the serving station.

Threshold refinement (TR) is now made possible by partitioned service (PS), see figure 3.5. In contrast to normal operation, the serving station does not remove the complete packet after having fulfilled the service. Instead, the packet is partitioned into identical pieces with respect to its service time. The simulation parameter $g$ is the granularity which determines the number of pieces into which the packets are divided. The number of events necessary for the service of a single packet is increased from a single event for the end of the service time in non-PS simulations to $g$ in PS simulations, one event for each partial removal of a packet in service.

To prevent PS from influencing the behaviour of the other nodes in the network model, e. g., the packet inter-arrival times at the other nodes, the departure event takes place at the same time as it would do in non-PS simulations. This means, the succeeding nodes are not informed about the partial end of the service of the packet before the service time has completely elapsed. This guarantees that the behaviour of the simulation itself remains unchanged and only the evaluation process for the arrival occupancy and by that the simulation control with RESTART is affected.

PS introduces an overhead into the simulation process which mainly results from the fact that more information has to be computed and stored in transition states. Furthermore, a higher number of events is needed for the completion of the service of a packet.

### 3.4.2 Non-RESTART simulations

To show the effect of PS, figures 3.6 and 3.7 show non-RESTART simulations. In figure 3.6, a simple M/M/1 model is used with a load of $\eta = 0.1$ in sub-figure 3.6 a) resp. $\eta = 0.2$ in sub-figure 3.6 b). Figure 3.7 shows simulation results for a simple M/D/1 model with a load of $\eta = 0.2$ in sub-figure 3.7 a) resp. $\eta = 0.4$ in sub-figure 3.7 b).

Most important is that the results of the simulations with the different granularities $g$, of which $g = 1$ means a non-PS simulation, perfectly match at the realistic $x$-values (the natural numbers in this case) and only differ at intermediate $x$-values. This applies for all simulations of figures 3.6 and 3.7.

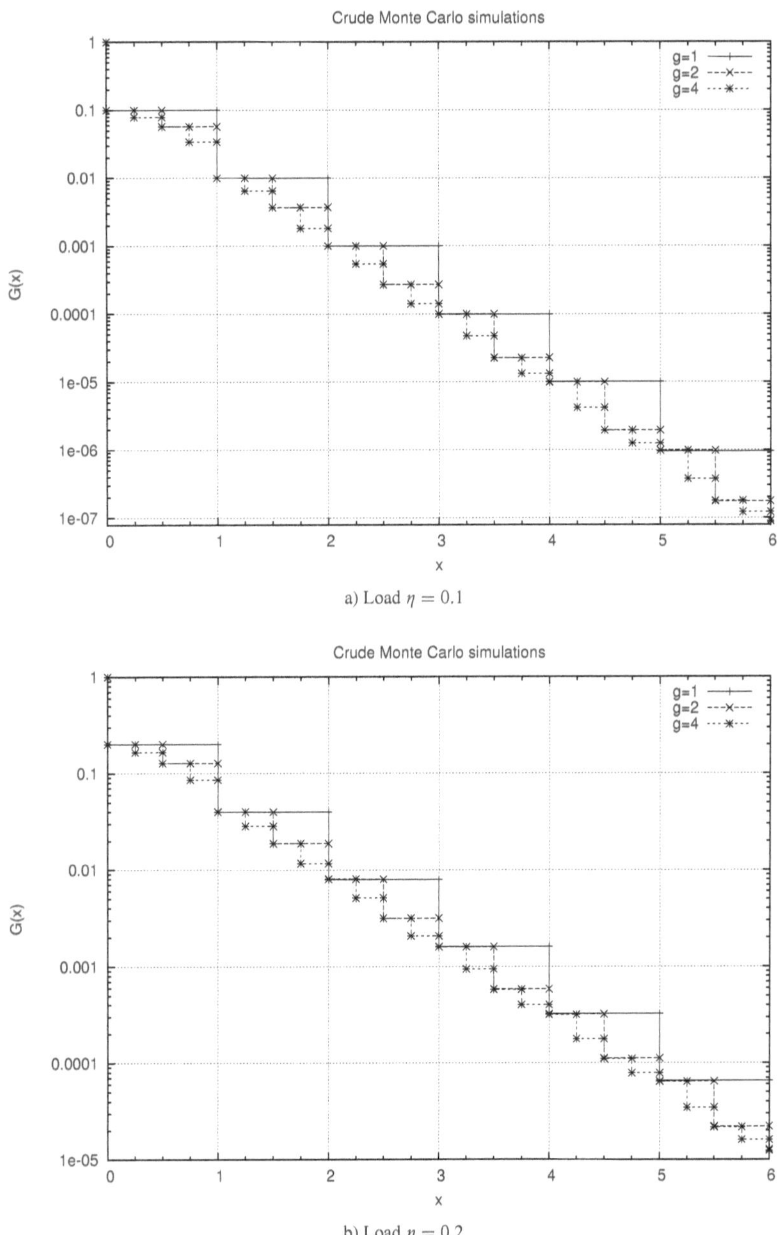


a) Load $\eta = 0.1$



b) Load $\eta = 0.2$

**Figure 3.6:** M/M/1 non-RESTART simulation

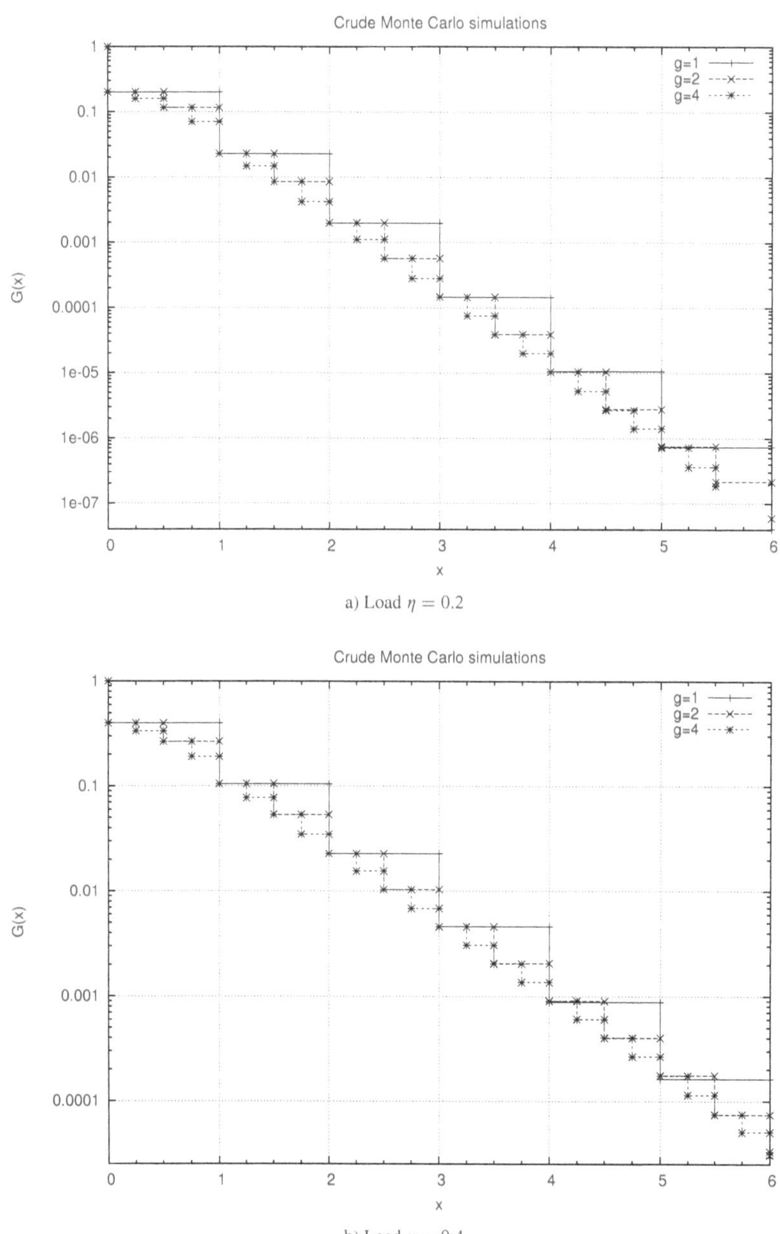


a) Load $\eta = 0.2$

b) Load $\eta = 0.4$

**Figure 3.7:** M/D/1 non-RESTART simulation

| M/M/1 | | |
|---|---|---|
| $\eta$ | $g$ | samples/sec |
| 0.1 | 1 | 210684 |
| | 2 | 140619 |
| | 4 | 112872 |
| 0.2 | 1 | 176465 |
| | 2 | 98679 |
| | 4 | 94804 |

| M/D/1 | | |
|---|---|---|
| $\eta$ | $g$ | samples/sec |
| 0.2 | 1 | 203810 |
| | 2 | 139617 |
| | 4 | 105779 |
| 0.4 | 1 | 209146 |
| | 2 | 125159 |
| | 4 | 103233 |

maximum relative error of 5 %
evaluation between $x = 0$ and $x = 20$

**Table 3.1:** Simulation overhead of PS simulations

Table 3.1 shows the simulation overhead of partitioned service. It results from the higher number of events which are needed to finish a packet in service, depending on the granularity. It is shown for all graphs the number of samples which the simulation has been able to produce per time period (seconds). The greatest performance decrease takes place between non-PS simulations and PS simulations with the smallest possible granularity $g = 2$. Further performance decrease for higher granularities is much smaller. The performance decrease has to be taken into consideration when applying PS to RESTART simulations. Too high granularities will have no more gain in efficiency for RESTART, but they will have further loss in efficiency for PS instead.

### 3.4.3 RESTART simulations

Figure 3.8 shows the RESTART simulations for both the M/M/1 and the M/D/1 model, but with lower load $\eta$ to stress the capabilities of RESTART and to chose configurations for which threshold refinement leads to a significant gain in efficiency. Table 3.2 shows the performance results for these simulations.

For the M/M/1 simulations in figure 3.8 a) and the upper part of table 3.2, the load $\eta$ has been chosen such that the optimum thresholds have a distance of $\Delta I_i = I_i - I_{i-1} = 0.5$ for $\eta = e^{-4}$. This leads at first sight to an optimum granularity of $g = 2$. The results in the table, however, show a granularity of $g = 4$ to be even more efficient although only one additional step has been introduced. The reason for this is the special situation at $x = 0$. For both $g = 2$ and $g = 4$, all thresholds from $I_1$ to the end have an optimum distance to their neighbour thresholds, and all steps needed comparable simulation times to finish.

For the first threshold $I_0$ the fact that $G(0) = \mathrm{P}\{X > 0\} < 1$ leads to a different distance between the corresponding probability values of the minimum value of the value space (which is 0) and the first threshold $I_0$, namely between $\mathrm{P}\{X \geq 0\} = 1$ and $H_0 = G(I_0)$. Equation (3.15) applies:

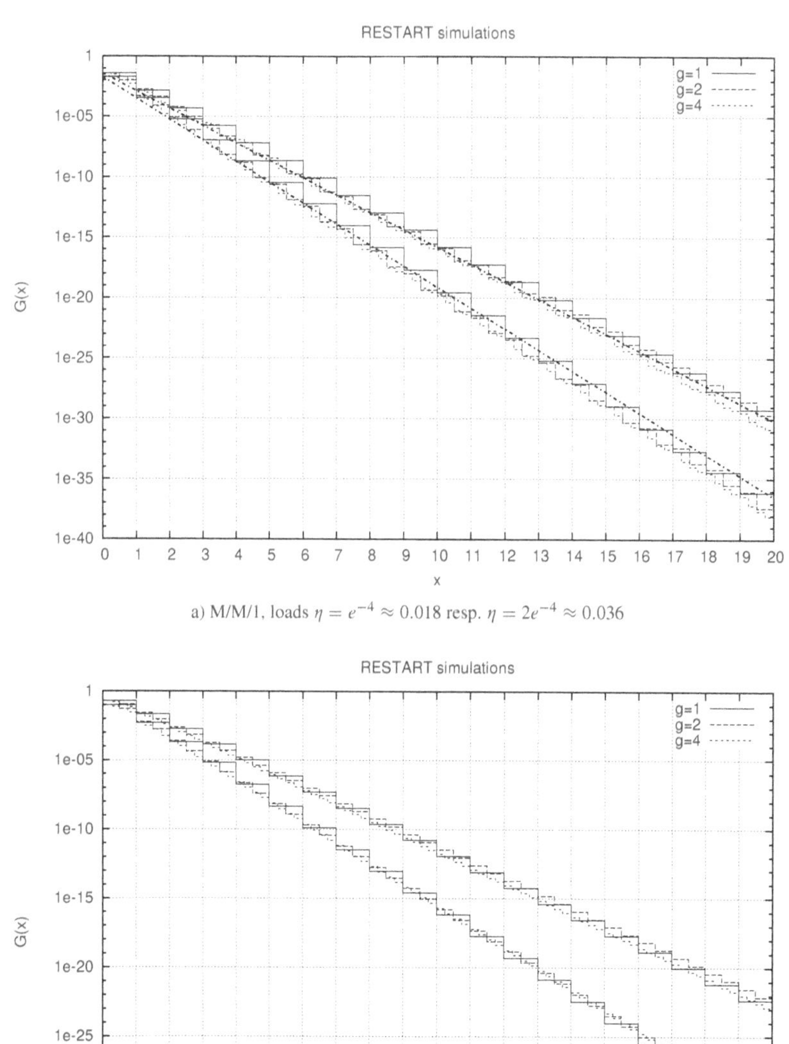


a) M/M/1, loads $\eta = e^{-4} \approx 0.018$ resp. $\eta = 2e^{-4} \approx 0.036$



b) M/D/1, loads $\eta = 0.1$ resp. $\eta = 0.2$

**Figure 3.8:** PS RESTART simulations

| M/M/1 | | | | | |
|---|---|---|---|---|---|
| $\eta$ | $g$ | steps | samples | time [sec] | speed-up |
| $e^{-4}$ | 1 | 19 | 704922000 | 12330 | *1.000* |
| | 2 | 39 | 50022006 | 1767 | 6.978 |
| | 4 | 40 | 23550018 | 1071 | 11.513 |
| $2 \cdot e^{-4}$ | 1 | 19 | 147288000 | 2727 | *1.000* |
| | 2 | 39 | 21513013 | 782 | 3.487 |
| | 4 | 42 | 17445021 | 876 | 3.113 |
| **M/D/1** | | | | | |
| $\eta$ | $g$ | steps | samples | time [sec] | speed-up |
| 0.1 | 1 | 19 | 236761002 | 3624 | *1.000* |
| | 2 | 39 | 25765004 | 775 | 4.676 |
| | 4 | 40 | 11536018 | 486 | 7.457 |
| 0.2 | 1 | 19 | 45892001 | 630 | *1.000* |
| | 2 | 39 | 11294011 | 332 | 1.898 |
| | 4 | 30 | 5842149 | 185 | 3.405 |

maximum total relative error of 5 %

**Table 3.2:** Efficiency of PS in RESTART simulations

$$H_{0|X\geq 0} = \frac{H_0}{1} = G(I_0) \tag{3.15}$$

For a load of $\eta = e^{-4}$, this distance is $G(I_0) = e^{-4(I_0+1)}$, for $\eta = 2e^{-4}$ the distance is $G(I_0) = (2e^{-4})^{I_0+1}$. This means, the first threshold should be set to the smallest possible value, which is $I_0 = 0.5$ for $g = 2$ and $I_0 = 0.25$ for $g = 4$. The first step in the interval $[0, I_0]$ occupies in such cases the greatest portion of the total simulation time. This is why the efficiency for $g = 4$ is significantly higher than for $g = 2$ in the case of the lower load $\eta = e^{-4}$. For the higher load, the distance $G(I_0)$ for the first possible threshold $I_0$ is closer to the optimum than for the lower load case, which leads to a smaller reduction of the sample size for $g = 4$ compared to $g = 2$. This reduction is too small to compensate the higher computation effort for the higher granularity, which results in lower total efficiency.

The same statements can be made for the M/D/1 model, figure 3.8 b) and the lower part of table 3.2. For the lower load $\eta = 0.1$, the optimum $\Delta I_i \approx 0.58$ which is close to 0.5. This leads to a behaviour comparable to one for the lower load of the M/M/1 model. For the higher load $\eta = 0.2$ it is $\Delta I_i \approx 0.724$ which is close to 0.75. For $g = 1$ and $g = 2$, the possible $\Delta I_i$ are quite far away from the optimum $\Delta I_i$, but with a granularity of $g = 4$ the thresholds can be set very close to the optimum ones. This explains the increase in efficiency for $g = 4$ and $\eta = 0.2$ of the M/D/1 model even with only 30 steps (=29 thresholds) which are set at optimum positions.

### 3.4.4 Conclusions on threshold refinement

Threshold refinement supports the placement of the thresholds closer to the optimum. This is important on the one hand for simulations of random variables of which the lowest possible value in the value space has a high probability. It results in a jump in the distribution function, and dimensioning the first step as small as possible is required in that case. This can be reached by a high granularity with the drawback of higher computation effort. On the other hand, distribution functions which are very steep in some intervals or everywhere need threshold refinement to increase the conditional probabilities of reaching a threshold from the next lower one. The optimum granularity depends on the jump at the lowest value and on the predicted (local) steepness of the distribution function.

Figures 3.6 and 3.7 show how accurately the distribution function $G(x)$ is represented by the PS simulations at all possibles values of $x$. Further, figure 3.8 shows the RESTART simulations. In the M/D/1 case, figure 3.8 b), the agreement of the PS simulations ($g > 1$) with the non-PS simulation ($g = 1$) is very good, and it is even better in the M/M/1 case, figure 3.8 a). Considering the achievable speed-up (table 3.2), threshold refinement can contribute well to the overall simulation accuracy.

# Chapter 4

# Distributed RESTART

For certain models, RESTART is a powerful method to simulate events up to an extremely low probability with sufficient accuracy and within reasonable time. As shown in chapter 3, the speed-up can be very high. Generally, parallelisation can speed-up simulations, depending on the problem to solve, see section 2.3. It appears natural to consider the possibilities to combine these approaches and to find out which way of parallelisation provides optimum additional speed-up for RESTART simulations.

It is essential to note that the parallelisation approach is not applied to a straightforward simulation method but on the RESTART simulation speed-up method. This affects mainly the way the decomposition of the simulation procedure is done, as it will be shown in more detail in section 4.1.1.

In the following it is shown which aspects have been considered for the development of distributed RESTART. The simulation toolkit MuSICS mentioned in chapter 3 has been extended with the implementation to produce the simulation results presented in this chapter. More details are found in [LG02] and [LG04].

## 4.1 Methodological aspects

On the way to an appropriate method for parallelising RESTART, several methodological aspects have to be considered. The choices which are possible for each of these aspects are discussed, and it is explained which choice is appropriate for MuSICS. The most relevant attributes of these possible choices are the scalability and the communication overhead.

### 4.1.1 Decomposition

Different approaches regarding the principal decomposition methods are introduced in section 2.3. Here, they are discussed with respect to their applicability to RESTART.

#### 4.1.1.1 Model division

The model division is introduced in section 2.3.3. For this approach, model components have to be identified to decide about the way they will be assigned to processes. These components need to be as autonomous as possible to minimise the necessary communication between the processes. Components can be the network nodes, preferably including the associated queue. This has the advantage of eliminating necessary communication between queue and node about the transfer of a job from the queue to the node. Only the transfer of jobs from nodes to other nodes have to be communicated.

A significant limitation of the model division is the scalability. For small network models with one source and one or two servers, the simulation could only be divided into two or three processes. Assigning the queues to separate processes would only slightly increase the scalability at the expense of the disadvantage discussed above.

Apart from this and the other usual problems with this approach discussed in section 2.3.3, the RESTART-specific actions have to be initiated and controlled. The control of the RESTART/LRE procedure is bound to the evaluation of a single random variable. The process of the network component which provides the values for the random variable has to perform the evaluation. It has to initiate the store and restore operations on all involved simulation objects on all processes.

The main aspect which disqualifies this approach for MuSICS is the very frequent transfer of job related data between the processes representing the network nodes.

#### 4.1.1.2 Functional and load division

Functional division usually means to divide simulations into functions like evaluation, random number generation and the simulation itself, as introduced in section 2.3.1. A different extended kind of functional division is considered for application with RESTART.

Apart from the possibility to separate the above mentioned functionalities, the simulation functionality can be divided into trajectories. A trajectory or trace is a sequence of simulated values starting from a restored state which has been previously stored and ending at a situation at which a new state has to be restored. At the end of a trace, possibly the reached system state is stored again. A trace can be considered as a part of a sequential RESTART simulation run. The simulation of a trace is completely independent of the rest of the simulation which is why several traces can be simulated concurrently by different processes. This is rather a load division even though only simulation parts are distributed to separate processes.

**Evaluation**

Conforming with the original idea of functional division, one process has to be responsible for the evaluation. Irrespective of the way how the simulation results are merged, they have to build one central result. One way is that the processes doing the simulation of the traces perform some pre-evaluation of the results and calculate statistics which can be merged by the central evaluation instance. This requires that the applied evaluation method is capable of merging these statistics. The other way is simply to have the simulation processes send all observations to the central evaluation instance which evaluates all observations within a single procedure as if they came from a single sequential simulation.

In RESTART/LRE it is essential for the evaluation procedure to have access to the intermediate results since the evaluation procedure also controls the simulation. This means that the intermediate statistics provided by the LRE also include the current actual value of the relative error. On this parameter, an important decision for the simulation is based, namely whether a step is completely simulated regarding the target maximum relative error. Thus, a central instance deciding on the simulation status and the required actions needs to get results or statistics from the simulation processes at regular intervals, not necessarily periodically.

Further down in section 4.1.3, the centralised and decentralised evaluation approaches are compared in more detail.

### Work distribution

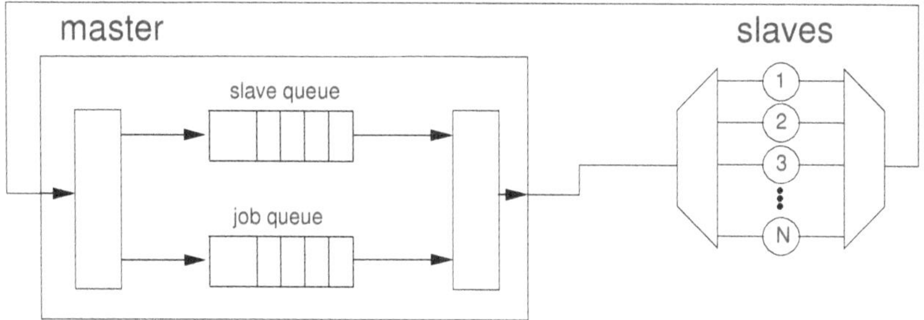


**Figure 4.1:** Master-slave architecture

Another central instance – the master process – has to manage the simulation work to be done. A single stored system state in RESTART will be used (replicated) $R_i$ times as the starting state of a new trace in step $i$. This leads to a number of $R_i$ new simulation jobs for each stored system state of the step $i-1$. These system states are extended with the information on the step it belongs to and by this form the simulation jobs. They are provided to simulation processes asking for new work after having completed the simulation of a single trace. The simulation processes are called slaves.

Such a central instance needs to know when a job has been simulated to keep the replication counter consistent. Thus, it is efficient to let this instance control the job distribution to the simulation processes. The alternative is to let a simulation process itself replicate a state and distribute the jobs to the other simulation processes. This raises the problem that it is not known to these processes which other process is still performing a simulation. Either each simulation process would need a queue for incoming work requests (job) which has to be managed by some other process, or the process sending the job would have to wait. The latter is prohibitively inefficient, and the first needs some non-blocking communication mechanism lacking the possibility of acknowledging the acceptance of a job.

A central instance managing the evaluation and by that knowing the overall simulation status can most efficiently manage the work distribution. Only the central manager has to know about the number and status of the simulation processes. It can use queues to manage the available jobs and the slaves waiting for a new job to simulate. This is shown in figure 4.1.

**Simulation**

In addition to the auxiliary functionalities described above, the main work is assigned to processes doing the actual simulation, all containing the complete simulation model.

Combination of the above functionalities are possible. One is the combination of a central master evaluation process with a decentralised pre-evaluations, see section 4.1.3.

### 4.1.2 Communication

Processes that need to exchange information have to communicate the information in some way. One process is willing to send information, and another process expects to receive information. If the receiving process can be addressed, the other process can explicitly send the information to it. Another way for a process willing to send is to announce the availability of information, and other processes which are interested in this information can collect it without the sending process knowing the receiving process.

#### 4.1.2.1 High Level Architecture

In the High Level Architecture (HLA) [KWD99] [Fuj00], an implicit communication is used. In the explicit communication, the process representing the information source directly addresses the process representing the information sink. In contrast to that, within the HLA, the information source process only makes the information available and announces the availability. The information sink process has to check for the availability of the information it wants to receive. In case of a positive check, the process has to fetch the information on its own. Generally, the source does not know which sink, if any, has collected the information, and vice versa.

To make this kind of communication possible, some central instance is needed which is known to all communicating processes, since otherwise the direct addressing of the peer process would still be needed. The so-called Run-time Infrastructure (RTI) undertakes this task. Processes needing information have to tell the RTI which type of information they need. They *register* for this information. The RTI is now able to tell the process about newly available information that is relevant for the process, every time the status of availability changes. The processes providing information have to tell the RTI the type of information, and they have to announce changes to the availability status to it. The RTI can forward this announcements to the processes interested in the corresponding types of information.

This type of parallelising simulations has been invented by the Department of Defence (DoD) for a special type of simulation, the distributed interactive simulation (DIS). This is a special type of real-time simulation which is distributed to a number of users interactively operating their own simulators. With the HLA, even heterogeneous simulators can be coupled.

In contrast to the real-time simulations, the simulations considered in this thesis represent the class of so-called as-fast-as-possible (AFAP) simulations. In [LKG01], it has been identified that in the application of the HLA on this class of simulations the management of the model time is a critical issue.

On the other hand, parallelising RESTART with the HLA would lead to a master-slave architecture similar to figure 4.1 with independent slaves and only jobs and results to be communicated. Nevertheless, the communication needs to be as fast as possible (non real-time). Since the HLA is quite complex and has been designed for real-time simulations, the performance of the implementations has been found to be insufficient for the parallelisation of RESTART with its very frequent necessity to exchange information.

#### 4.1.2.2 Message passing

The explicit communication is called message passing and can be implemented according to the message passing interface (MPI) [GLS99]. Each single communication (a message) has one sender and one or many receivers, and the sender knows the receiver(s) of the message and vice versa.

The MPI is an application programming interface (API) which is widely used by many implementations for many architectures and platforms, and it has been designed for efficient communication. Thus, for most systems which are used for distributed computing, optimised implementations are available. For the simulations of this thesis, different MPI implementations have been used, depending on the platform.

### 4.1.3 Merging statistics

In the central evaluation, all obtained values have to be communicated to the evaluation process. This leads to a large communication effort. An alternative approach is to have the simulation processes evaluate their own samples and send the result of the evaluation to the master evaluation process which combines the results to an overall result.

A prerequisite is to keep necessary auxiliary values for the result combination procedure. If, e. g., only the overall mean value is needed, sending only the mean values of the simulation processes to the master is not sufficient if the sample sizes are different. For evaluation of distribution functions as done with the LRE, the intervals must be at the same locations for all participating simulation processes. See also section 3.2.3.

In the following consideration, it is assumed to have a central process – the master – calculating the overall result, and a number of $M$ simulation processes – the slaves, all with the same configuration evaluating the same value space. The statistics are done with the LRE method.

#### 4.1.3.1 Simulation

The pre-evaluated results of the slaves are sent to the master process. Since the intervals of the LRE objects are the same for all slaves, the result combination, or the merging, is done by cumulating all internal counters of all contributing LRE objects for each interval. The main counter represents the number of occurrences of the interval value, and another counter manages the leftward transitions according to the corresponding 2-node Markov chain in figure 2.8.

In the centralised evaluation, all observations of a certain slave are sent to the master after a simulation job has been finished. The values collected during the simulation of a single job are called a trace. These traces can be grouped to form trace groups of, e. g., of 1000 values, in order to reduce the number of messages. It is more efficient to have a smaller number of messages to be sent than a large number with the same cumulated message size, see [LG02]. There, it has been further pointed out that large trace groups cause wasting of simulation time. It is possible that the master has already collected enough values and has reached the target error level, but some slaves are still simulating since they have not yet filled their trace group. The main reason for this to happen is the nescience of the slaves about the current error level.

In the decentralised approach, the master does not know about the error level reached by a certain slave. Instead, the slaves do know about this, but only about their individual error level, not about the overall error level. The total error of the merged statistics for step $i$ is shown in equation (4.1) where $n_{i,k}$ represents for each slave $k$ the number of observations obtained for step $i$.

$$d_i \propto \frac{1}{\sqrt{\sum_{k=1}^{M} n_{i,k}}} \tag{4.1}$$

The problem that the master does not know about the error level at the slaves and by that also about the current overall error has to be solved. Some instance has to decide about the finishing of step $i$. With the above equation, it is possible to have the slaves themselves decide about the finish of their individual contribution to step $i$. Under homogeneous circumstances, all slaves can be assumed to produce comparable numbers of observations within the same time. For $M$ slaves, each slave produces approximately an $M^{th}$ part of the overall needed observations $n_i$ to reach the total error $d_i$ for step $i$, thus $n_{i,k}$ is as follows:

$$n_{i,k} = \frac{n_i}{M} \qquad \forall\, k = 1 \dots M \tag{4.2}$$

Using equations (4.1) and (4.2), the individual error level $d_{i,k}$ is

$$d_{i,k} = d_i \cdot \sqrt{M} \tag{4.3}$$

With these adapted error limits for the slaves, all slaves can autonomously perform their work and send afterwards the pre-evaluated results to the master process for merging. Probably, the achieved total error level $\tilde{d}_i$ will not be exactly $d_{i,k}/\sqrt{M}$. A significant oversampling leading to an actual relative error far below the target would waste simulation effort. This case is very unlikely, however. A small deviation of the target error level can happen, and in this case, the master initiates some more simulation work until the step $i$ is finally simulated sufficiently.

#### 4.1.3.2 Communication

Now it has to be considered the amount of data to be transmitted for an object containing the status of an LRE evaluation, or in other words the data needed to unambiguously reconstruct the corresponding LRE object on the side of the master process. If the target for the total error is reached without the master needing to distribute additional jobs after having received the pre-evaluated results from all slaves, this transmission is only needed once per slave and step $i$.

In the implementation of the LRE for the investigated computer architectures, an LRE object has a fixed part of about 200 byte and a value space dependent part with an amount of 20 byte per possible value resp. per interval. The latter is an array of a data structure consisting of a pointer, the value itself, and two counters for the occurrence frequency and the leftward transitions.

As an example, and without loss of generality, a discrete value space $x_i \in 0, 1, \dots, B$ is chosen for the random variable $X$. This leads to $B + 1$ elements in the array and to a overall message size $N_b(\text{decentral})$

$$N_b(\text{decentral}) = 200 + 20 \cdot (B + 1) \tag{4.4}$$

In the central approach, the total message size is compared to the decentralised approach as follows:

$$N_b(\text{central}) = 8 \cdot n_{i,k} \approx 8 \cdot \frac{n_i}{M} \tag{4.5}$$

Here, it is assumed to have values of double precision requiring 8 byte per value. While in the decentralised approach only a single message has to be sent, and possibly a few additional ones if the target error level has not been reached, in the central approach, a usually large number of values $n_i$ need to be obtained and sent. Even for quite large trace groups, the number of messages will be large.

Again considering the total amount of data to be transmitted to the master, it will be much larger for the central approach since $\frac{n_i}{M}$ will usually be much larger than $B$. Let, e. g., $B = 89$, $M = 4$, and a even quite low figure $n_i = 10^6$. For this example, in the central approach a factor of 1000 times more data has to be transmitted than for the decentralised approach. The decentralised approach, however, has not yet been implemented since major recoding of the LRE would have been necessary.

### 4.1.4 Decision

The decided method is the master slave topology with message passing using the MPI. The master manages the simulation work and distributes it to the simulation processes called slaves. The slaves send their results and newly found transition states to the master, and the master performs the evaluation centrally.

## 4.2 Simulations

A collection of simulation results is presented here to show the performance of the method chosen in section 4.1.4. Simulations have been conducted on different platforms and network topologies.

### 4.2.1 Platforms

The platforms on which the simulation has been conducted were a plain network of workstations (NOW), an SMP (symmetric multiprocessor) machine, and a combination of this. The latter was a network of dual processor machines.

#### 4.2.1.1 Network of workstations

As pointed out in [LG02], a NOW is easy to set up in the sense that cheap off-the-shelf hardware can be used to build a parallel computer. The communication between the processes, however, has to take place over network hardware. This is slow compared to communication on the level of the system bus as in SMP system or even massively parallel processors (MPPs).

#### 4.2.1.2 Symmetric multiprocessor

SMP systems have the advantage that communication between processes on different processors of the system is an intra-system communication, and it can take place on the fast system bus level. Every external network hardware puts overhead to the communication compared to the intra-system communication.

The main disadvantage is the limited scalability. SMP system with a high number of processors are either not available or at least comparably expensive.

#### 4.2.1.3 Network of SMPs

A combination of the two platforms above is to connect a number of SMP systems via external network hardware. The limited scalability of plain SMP systems does not apply, and small SMP systems with two processors are hardly more expensive than two single processor systems, they are usually even cheaper. Even four-processor systems are affordable.

The advantage of fast intra-system communication for neighbouring processes makes it a faster system than a NOW of single-processor machines, at for certain interaction structures, see section 2.3.

The structure of the chosen method with a master process and several slaves, however, does not benefit from this. The interaction structure is a star, and only the slave process(es) on the same machine as the master process can make use of the faster intra-system communication.

### 4.2.2 Network technology

Different network technologies are possible, of which only the well known Ethernet is considered, and furthermore a representative for high-speed networks, namely SCI (scalable coherent interface), on which simulations have been conducted in this thesis.

#### 4.2.2.1 Fast Ethernet

Ethernet is the most used network technology, it is cheap, and it is nowadays installed on-board without needing an extra network card. Fast Ethernet with 100 Mbit/s already is standard, faster transfer rates are common.

Even with high transfer rates, or throughput, however, the main disadvantage for high speed parallel computing with significant communication effort is the high latency of the Ethernet technology.

#### 4.2.2.2 Scalable coherent interface

As latency is critically important for efficient distributed computing, SCI as a technology with low latency has been considered. The transfer time of small messages from one machine to another is about 150 times less than with Ethernet. The throughput is higher than Gigabit Ethernet, but with the high number of small messages generated by the method discussed here, as identified in [LG02], the latency is the limiting factor.

The topologies possible with SCI are a one-dimensional ring, and grids of two and three dimensions. The setup used for the SCI simulations in this thesis consists of four dual-processor SMP machines (nodes), and thus, a ring topology has been used. For an increasing number of nodes, the number of dimensions has to increase to keep the communication efficient.

### 4.2.3 Simulation results

To evaluate the performance of the chosen parallelisation method, two comparisons are presented in the following.

#### 4.2.3.1 Plain NOW vs. SMP system

In this section, simulation runs are compared which have been conducted on a plain NOW and on an SMP machine. The SMP machine was a Sun Enterprise server with four processors with 400 MHz each. The NOW contains single-processor machines with 1 GHz Athlon processors connected via a Fast Ethernet network. To keep the simulation results of the NOW comparable with the SMP results, only up to (the quite small number of) 3 slaves have been used also in the NOW simulations.

| model | M/M/1 | | | | Jackson | | | |
|---|---|---|---|---|---|---|---|---|
| | time [s] | gain | eff (sl) | eff (tot) | time [s] | gain | eff (sl) | eff (tot) |
| single | 45.5 | 1.00 | 1.000 | 1.000 | 32.1 | 1.00 | 1.000 | 1.000 |
| 2 slaves | 48.7 | 0.93 | 0.467 | 0.311 | 33.6 | 0.96 | 0.478 | 0.318 |
| 3 slaves | 49.2 | 0.92 | 0.308 | 0.231 | 30.6 | 1.05 | 0.350 | 0.262 |

**Table 4.1:** Results for Fast Ethernet plain NOW

| model | M/M/1 | | | | Jackson | | | |
|---|---|---|---|---|---|---|---|---|
| | time [s] | gain | eff (sl) | eff (tot) | time [s] | gain | eff (sl) | eff (tot) |
| single | 136.5 | 1.00 | 1.000 | 1.000 | 107.0 | 1.00 | 1.000 | 1.000 |
| 2 slaves | 97.4 | 1.40 | 0.701 | 0.467 | 76.2 | 1.40 | 0.702 | 0.468 |
| 3 slaves | 74.3 | 1.84 | 0.612 | 0.459 | 53.9 | 1.99 | 0.662 | 0.496 |

**Table 4.2:** Results for SMP system

Tables 4.1 and 4.2 show the run-times of simulations with two different models. The first model is an M/M/1 model, the second one is a simple Jackson network with two sources and three stations. In these tables, the keywords have the following meanings:

**time:** Run-time in seconds

**slaves:** Number of slave processes. In the NOW case, this is equal to the number of slave nodes since single-processor machines have been used.

**single:** The non-parallelised simulation as reference.
**gain:** Ratio of the run-times of the non-parallelised simulation and the parallelised simulation. As reference value, the gain of the non-parallelised simulation is 1.0.
**eff (sl):** Slave efficiency, i. e., gain divided by number of participating slaves.
**eff (tot):** Total efficiency, i. e., gain divided by number of participating processes, which is the number of slaves plus one.

Originally, the different models have been chosen to investigate the impact of different message sizes on the performance. The run-times, however, result from optimised simulations where the slaves send groups of found transition states to the master and in turn receive a group of jobs to simulate from the master. This reduces the number of messages to be sent significantly, and, which is relevant for this consideration, it makes the communication overhead much less dependent on the size of the system states. The different run-times for the two models result from the simulation of the models, not from the different size of the communicated messages.

In the NOW simulations of table 4.1, the gain of the parallelised simulations is around 1.0 for all runs. That means, it is actually a loss in most cases. The efficiency values are low, and the slave efficiency values as well as the total efficiency values are lower in the simulations with the higher number of slaves. For both models, the behaviour is very similar.

In the SMP simulations of table 4.2, the gain of the parallelised simulations is higher than 1.0, and thus, there is indeed a gain in the application of the distributed RESTART on an SMP system. The slave efficiency only slightly decreases with an increasing number of slaves, and the total efficiency remains almost constant or even increases.

#### 4.2.3.2 Fast Ethernet vs. SCI technology

In this section, an M/M/1 model has been simulated on two systems with different network technologies. In the simulations of table 4.3, the inter-system communication has taken place over a Fast Ethernet network, and for table 4.4, the SCI network has been used.

As in the previous section, the tables show the run-times of the simulations with the resulting gain and efficiency values. Additionally, to achieve a higher precision for the gain and efficiency values, the number of observations simulated and by this the number of observations per second have been considered. The reference values are here the number of observations per second of the non-parallelised simulation runs.

In these tables, additional to the tables in section 4.2.3.1, the keywords have the following meanings:

**nodes:** Number of contributing nodes including node executing the master process. Since dual-processor machines have been used, this information is needed additionally to the number of (slave-)processes.
**observations:** Number of observations simulated, also called the sample size.

| slaves | nodes | time [s] | observations | observ./s | gain | eff (sl) | eff (total) |
|---|---|---|---|---|---|---|---|
| single | 1 | 237.47 | 62388774 | 262722 | 1.00 | 1.000 | 1.000 |
| 2 | 1** | 353.49 | 62379731 | 176470 | 0.67 | 0.336 | 0.336 |
| 2 | 2* | 264.20 | 62378458 | 236107 | 0.90 | 0.449 | 0.300 |
| 2 | 2 | 399.20 | 62306763 | 156080 | 0.59 | 0.297 | 0.198 |
| 2 | 3 | 303.82 | 62264142 | 204940 | 0.78 | 0.390 | 0.260 |
| 3 | 2* | 232.98 | 62182673 | 266901 | 1.02 | 0.339 | 0.254 |
| 3 | 3 | 273.09 | 62161802 | 227621 | 0.87 | 0.289 | 0.217 |
| 3 | 4 | 232.44 | 62111685 | 267215 | 1.02 | 0.339 | 0.254 |
| 4 | 3* | 190.16 | 62342438 | 327845 | 1.25 | 0.312 | 0.250 |
| 4 | 3 | 252.58 | 62325538 | 246753 | 0.94 | 0.235 | 0.188 |
| 4 | 4* | 173.05 | 62203616 | 359462 | 1.37 | 0.342 | 0.274 |
| 4 | 4 | 226.75 | 62350396 | 274979 | 1.05 | 0.262 | 0.209 |
| 5 | 3* | 190.65 | 62208029 | 326296 | 1.24 | 0.248 | 0.207 |
| 5 | 4* | 175.26 | 62254812 | 355215 | 1.35 | 0.270 | 0.225 |
| 5 | 4 | 217.94 | 62364940 | 286159 | 1.09 | 0.218 | 0.182 |
| 6 | 4* | 175.07 | 62229016 | 355442 | 1.35 | 0.225 | 0.193 |
| 6 | 4 | 217.15 | 62286768 | 286841 | 1.09 | 0.182 | 0.156 |
| 7 | 4* | 182.00 | 62377945 | 342733 | 1.30 | 0.186 | 0.163 |
| 8 | 4** | 180.46 | 62306823 | 345271 | 1.31 | 0.188 | 0.164 |

**Table 4.3:** Results for Fast Ethernet cluster

| slaves | nodes | time [s] | observations | observ./s | gain | eff (sl) | eff (total) |
|---|---|---|---|---|---|---|---|
| single | 1 | 237.47 | 62388774 | 262722 | 1.00 | 1.000 | 1.000 |
| 2 | 1** | 1035.65 | 62316546 | 60171 | 0.23 | 0.115 | 0.115 |
| 2 | 2* | 277.28 | 62322268 | 224762 | 0.86 | 0.428 | 0.285 |
| 2 | 2 | 283.88 | 62273311 | 219367 | 0.83 | 0.417 | 0.278 |
| 2 | 3 | 279.94 | 62266611 | 222431 | 0.85 | 0.423 | 0.282 |
| 3 | 2* | 199.19 | 62150887 | 312013 | 1.18 | 0.396 | 0.297 |
| 3 | 3 | 198.21 | 62203843 | 313822 | 1.19 | 0.398 | 0.299 |
| 3 | 4 | 197.35 | 62155652 | 314953 | 1.20 | 0.400 | 0.300 |
| 4 | 3* | 157.96 | 62207077 | 393814 | 1.50 | 0.375 | 0.300 |
| 4 | 3 | 157.90 | 62237533 | 394166 | 1.50 | 0.375 | 0.300 |
| 4 | 4* | 157.61 | 62326743 | 395448 | 1.51 | 0.376 | 0.301 |
| 4 | 4 | 158.78 | 62296409 | 392336 | 1.49 | 0.373 | 0.299 |
| 5 | 3* | 139.25 | 62343668 | 447704 | 1.70 | 0.341 | 0.284 |
| 5 | 4 | 138.92 | 62265749 | 448212 | 1.71 | 0.341 | 0.284 |
| 6 | 4* | 133.32 | 62231937 | 466795 | 1.78 | 0.296 | 0.254 |
| 6 | 4 | 133.72 | 62341118 | 466218 | 1.77 | 0.296 | 0.254 |
| 7 | 4* | 132.89 | 62373242 | 469362 | 1.79 | 0.255 | 0.223 |
| 8 | 4** | 323.85 | 62285788 | 192331 | 0.73 | 0.092 | 0.092 |

**Table 4.4:** Results for SCI cluster

**observ./s:** Number of observations simulated per second.

*: Node count values with this symbol mean that the node with the master process also holds one slave process.

**: Node count values with this symbol mean that the node with the master process also holds two slave processes, i. e., three processes on a dual-processor machine.

For the Fast Ethernet results in table 4.3, it can be seen an increasing gain with increasing number of participating slave processes. In the runs with a slave process sharing the node with the master process (with the * symbol), the gain is higher than in the other runs with the same number of processes. Obviously, the fact that a part of the communication can be done on intra-system level is essential. Among these runs, the ones with the higher number of nodes are faster. On a node with two slave processes communicating via the inter-system network with the master, twice the number of messages and twice the amount of traffic has to pass the high latency network connection than on a node with one idle processor.

In the SCI simulations, the gain is in almost all cases notably higher than in the corresponding Fast Ethernet simulation. In turn, this applies also for the efficiency values. It also has a certain gain already with only 2 slaves.

In contrast to the Ethernet simulations, the performance of the SCI simulations is completely independent of the distribution of the processes over the nodes. Exceptions are the cases when there is more than one process assigned to a processor, indicated with the ** symbol. This independence shows that the high speed network connections do not suffer from a higher number of messages. Furthermore, the CPU is hardly bothered at all with the network traffic. In the Ethernet case, an idle processor can take over the overhead CPU load that is due to the network traffic, while in the case of both processors simulating, the overhead CPU load is subtracted from the resources for the simulation processes.

## 4.3 Conclusions

A reduction of the time from the start to the end of a RESTART/LRE simulation is achievable by the application of the distributed approach. This is referred to as the gain. Depending on the platform and the network technology, a minimum number of slave processes is required to achieve an actual gain.

A common property of parallel computing is the fact that the speed-up compared to the non-parallelised case is non-linear and less than the number of participating processors. In other words, the efficiency will be smaller than 1. This applies also to the distributed RESTART approach.

The gain, if any, achieved with the Ethernet based system is very low, see tables 4.1 and 4.3. These systems cannot efficiently cope with the high communication effort that is mainly due to the high number of messages as identified in [LG02].

The system with the low latency SCI network technology shows increased efficiency compared to the Ethernet based systems. In this configuration, having available at least three dual-processor nodes makes the application of distributed RESTART reasonable, since the results will be available faster. The efficiency, however, is still quite low, and it is expensive in terms of hardware resources. The network hardware is much more costly than usual Ethernet equipment without increasing the efficiency significantly.

On SMP systems, the distributed RESTART behaves indeed as expected for parallel computing. A gain is achieved already with the smallest possible number of slave processes, and the efficiency values are very similar. Although these statements can only be made for a small setup with very few processes, it is remarkable that the efficiency values outperform all other efficiency values of the NOW and the cluster systems, even with the high-speed network technology.

The original intention has been to make a distributed RESTART approach applicable on inexpensive systems. This, indeed, is possible. Only on SMP systems, however, the approach provides a satisfying performance, and SMP systems with a high number of processors are expensive.

An approach for future investigations is to take a number of SMP systems with at least four processors each and set up an additional hierarchy level for the evaluation and the simulation control. This means each SMP machine has its own local master process, and all slave processes only communicate with their local master. The local masters have to synchronise the simulation progress information with a single global master process at certain intervals. Also for small scale SMP systems with two processors per node, however, as in the simulations presented in this thesis, this approach can be investigated. It will lead to insight about which hierarchical structure with which parametrisation will lead to performance improvements. A binary tree would, e. g., be an obvious choice for the small scale SMP system.

# Chapter 5

# Short-Term Dynamic Simulation Concept

In 3rd generation mobile radio networks like UMTS, network planning has changed from the way it has been for previous generations like GSM. In such systems, the Quality of Service (QoS) almost entirely depended on the location where the user is when accessing the service. This means, the parameters responsible for the radio coverage, mainly antenna positions, angles, and azimuths, can be planned without taking into account the traffic which can occur in the system.

GSM like systems are capacity limited [HT00], and the system capacity can be considered separately from the coverage which is independent of the traffic. This is why the system capacity can be considered after the antenna positions have been decided.

In interference limited systems like UMTS, the signal quality which determines the coverage is not independent of the traffic. Users share the radio frequencies with the use of orthogonal codes. The orthogonality is not perfect and decreases with the number of codes in use, and thus, with the number of users. As a result, knowledge about the coverage to decide on the antenna positions can only be gathered by simulations which also take into account predictions about the traffic.

Three distinct simulation concepts can be used in UMTS network planning, of which the new short-term dynamic simulation concept will be introduced and investigated. Starting with the basic concept and an analytical investigation in this chapter, details regarding the simulation of UMTS scenarios and the simulation toolkit are discussed in chapter 6.

## 5.1 Classification

The different simulation approaches which are possible for network planning with the focus on UMTS radio networks are the *static*, the *dynamic*, and the *short-term dynamic* (STD) simulation method. They have already been shortly introduced and compared qualitatively in section 2.2.1 with respect to simulation speed-up aspects.

According to the discussion in section 2.2, the state space is infinite due to the large number of dimensions and the – in some cases – continuous value space within these dimensions. Consequently, stochastic simulation techniques are required.

In a stochastic simulation, the modelled system is represented by a stochastic process $\{X(t)|t \in T\}$. Since discrete event simulation is considered, $T$ denotes an index set for the simulated time span, over which $t$ varies as the index parameter denoting the current simulation time. A possible value of the value space of the RV $X(t)$ is called a *state* [Tri02], and the set of possible values for $X(t)$ is the *state space*.

In a *static* or snapshot simulation, a sequence of independent snapshots is generated and evaluated. A snapshot represents a possible system state of the simulated model. The steady state distribution function of all parameters which build a snapshot must be known in advance or has to be assumed. These parameters are the input parameters, and they can be the number and the positions of users in a scenario, probabilities of profiles for the users and the like. In the simulation, the generated snapshot is investigated with respect to interferences between objects in the scenario. Such interferences can lead – in the case of UMTS – to the calculation of load on the air interface, soft-handover regions, and much more, see [TPL+03].

With this type of simulation technique, in contrast to the other techniques discussed in this thesis, the system is not represented by a stochastic process. A time axis does not exist, thus, the investigated system states are all snapshots with no history data. No information about the past and the future of these snapshots is incorporated.

As a major aspect, Quality of Service (QoS) statistics like the probability that a user cannot establish a call because load limits are exceeded, can also be obtained by *static* simulations. Statistics which need some sort of system history, however, cannot be obtained. Such history data can be, e. g., the elapsed time of a call, the duration of a finished call, and positions of moving users in the past. Only with such knowledge about the history of the system or a part of it, it is possible to deduce statistics as the frequency of regularly finished calls and the probability of a call being dropped for a moving user. For these statistics, it is necessary that such system dynamics providing history data are included into the simulation.

Naturally, the need to include dynamic aspects into the simulation would lead to a decision in favour of a *dynamic* simulation. All desired dynamic aspects can be considered as detailed as needed. This is very useful, e. g., for investigations on protocols where the level of detail is needed to verify the accurate protocol behaviour. For complex systems and the requirement to cover the state space well, as it is the case for UMTS network planning, the disadvantages of the dynamic simulation method are shown in section 2.2.1.

The *short-term dynamic* (STD) method has been originally designed for UMTS network planning purposes, see [LPTG03]. It combines the faster state space coverage with the ability to consider dynamic aspects. The user dynamics used to evaluate further QoS aspects, compared to static simulations, are mainly the mobility (section 6.2.3) and the session of users (section 6.2.4). Important additional QoS aspects are real dropping, packet delays in packet switched (PS) services, up- and downgrading of active calls, and the number of soft-handover events or

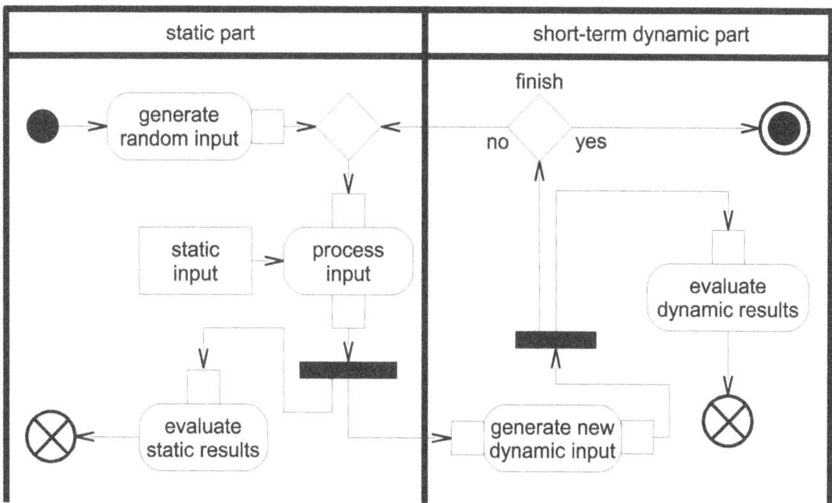


**Figure 5.1:** Activity diagram of a single STD window in STD simulation, UML 2.0

cell changes per time unit or per session. For detection of dropping, users who have previously started their session must be distinguished from newly emerging users. For the latter, a not accepted call is a blocked call. In static simulation, this distinction is only possible on a statistical basis.

## 5.2 Concept

### 5.2.1 Snapshot

For the short-term dynamic simulation concept, it is necessary, first of all, to be able to generate system states which follow a combination of stationary distributions of the involved objects and attributes of the state. A generated system state is a so-called snapshot. It consists mainly of a static input part, a random input part, and dynamic aspects which cause changes to the system state and which are triggered by events.

Figure 5.1 shows an activity diagram in UML 2.0 for a single STD window. The activity has two partitions, namely the *static part* and the *short-term dynamic part*. The components are explained in the following.

#### 5.2.1.1 Random input data

These are all the parts of a snapshot that follow stationary random variables. For network models, these are the number of present users, their positions, their state of activity, etc. In the simple model, this is only the queue occupancy. In case of a non-Markovian service process in the simple model, the elapsed service time of a currently served job would be another random part.

The random input has to be generated during the simulation. At the start of the STD window, indicated by the starting node (the small filled circle) of the activity diagram, the *generate random input* action generates this part of the initial snapshot. The square attached to the action symbol represents the generated data object which flows to the *process input* action. This square always represents this random part of the input data, also at all other action nodes in the diagram.

#### 5.2.1.2 Static input data

The *static input* rectangle is the object node representing all data, parameters, and information which do not change between subsequent snapshots. The static scenario data are of this category, like the positions of antennas, buildings, and other obstacles, and the clutter maps. Further, also the dynamic objects have static parameters, like power limits of mobile stations. In the simple model used for the analytical investigation in section 5.3, this is simply the fact that the model has one server and one queue.

Different from the *random input*, the *static input* is always present from the beginning of the main simulation. It only has to be retrieved once from some data source.

#### 5.2.1.3 Process input action

The action *process input* processes the *random input* of the initial snapshot as well as the dynamically changed data during the STD window. The latter data object is still of the type *random input*, only its generation is different.

Generally, interferences between objects which have been assigned parameters from the *random input*, like position and status, are considered here. This is necessary if the parameters of random objects are not independent of other random objects but have influence on them. In UMTS, e. g., the transmission power of mobile stations does not only depend on their position relative to base stations, but also on their positions relative to other mobile stations which can in turn change their activity status assigned by the *random input*.

In case of the simple model, the random input data is passed through unchanged to the evaluation and to the short-term dynamic part of the activity.

#### 5.2.1.4 New dynamic input data

The dynamic parts of the system do not occur in the snapshot itself. The action *generate new dynamic input* initiates the changes in the system state regarding the random parts, like changing the position of the users (mobility) and their activity state (session dynamics). These changes to the processed input are triggered by events. This takes place in the mentioned action which represents the actual event driven part of the simulation system. An STD window as a whole represents a small but complete discrete event simulation (DES).

In a more detailed diagram, another part of the *static input* data would be additional input for the *generate new dynamic input* action. These parts are the static parameters describing the random numbers for the events. E. g., a static matrix is needed which holds for every possible position the probabilities to keep the direction or turn to another possible direction when moving, see section 6.2.3.

#### 5.2.1.5 Miscellaneous

The *process input* action sends the processed input also to the evaluation action *evaluate static results*. Similarly, the dynamically changed input data is sent to the evaluation action *evaluate dynamic results*. The flow final nodes – represented by the circles with the "X" inside – after the evaluation actions make the activity diagram consistent. Actually, the evaluation results have to be stored somewhere, and also the evaluation actions need access to more information than is shown in the diagram. The evaluation is only indicated because the main focus in the diagram is on the generation and processing of the data building the system state.

The decision symbol named *finish* decides on whether the STD window can be terminated. The criterion can be based on real convergence measurement of obtained data, or it can simply be the expiration of a fix model time duration. The merge diamond will in the other case forward the dynamically changed data back to the *process input* action. Only once, at the beginning of the STD window, the merge symbol will be reached from the *generate random input* action which provides the random part of the initial snapshot.

### 5.2.2 STD windows

In the previous section, the concept has been described on the level of the system state, and it has been described what a single snapshot of the system consists of. In addition to this, figure 5.1 shows the processing of the system state data within an STD window with an orientation on the termination of the STD window. Figure 5.2 shows from another possible perspective, how the STD windows emerge.

On the $z$-axis, a parameter is plotted which represents the system state. The system state will usually be multi-dimensional, but to demonstrate the concept, a simple and general system is shown. On the $x$-axis, the simulation time is plotted.

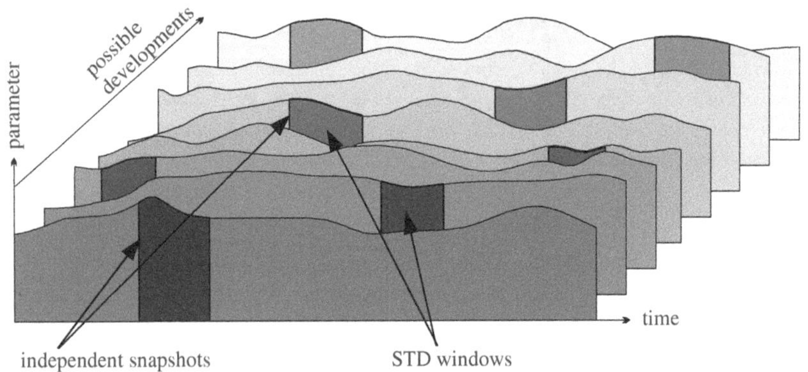


**Figure 5.2:** STD windows

A filled curve starting at the origin of the diagram stands for a possible development of the parameter. An *independent snapshot* is now taken at a random point in time. This corresponds to the snapshot which is generated by the *generate random input* action in figure 5.1.

Starting from this snapshot, the STD window is simulated. In the figure, the STD windows are the highlighted areas on the filled curves. After the STD window is finished, another independent snapshot is generated by chosing a random time from another possible development of the system parameters.

The image of different possible simulation developments on the $y$-axis is used to express two properties of the STD simulation concept. First, not only a single development starting from some reference state like an empty system is possible. Instead, the model usually has an infinite state space (see section 2.2), and from each possible system state a large or even infinite number of possible states can occur as the subsequent state after an event has arrived.

Secondly, the idea of independent snapshots is emphasised. From the perspective represented by figure 5.2, an STD window is an extract of a possible simulation development, and from these as many are considered as STD windows are needed. If only a single very long simulation development would be used to take the STD windows from, the fact that there is no long term correlation between the initial snapshots of subsequent STD windows would not be pointed out sufficiently. The independence of subsequent initial snapshots, however, is one of the most fundamental assumptions in the application of the STD simulation concept.

## 5.3 Analytical investigation

To evaluate the short-term dynamic simulation method, it has to be investigated analytically. It is important to look at the behaviour of STD simulations for very basic models to understand the concept and the benefit of the STD method. Only by applying such models, it is possible to use analytical methods to calculate the behaviour.

The application of the simulation method on certain domains and scenarios, however, has to be justified. Thus, the simple model selected for the analytical investigation has to represent essential basic properties of the simulation model. The service which will remain one of the most important ones for the mobile radio networks, especially in the early phases of the network use, will be the speech service. The speech model used in all simulations was configured to have the duration of its on- and off-phases negative exponentially distributed, in both up- and downlink. Aggregation of user sessions with this behaviour leads to a Poisson process which is a counting process having the Markov property.

The most basic model is the M/M/1 model. With its Markov property it allows for important simplifications. The distribution of the arrival resp. departure time of the next arrival resp. departure does not depend on the time elapsed since the last arrival resp. departure has taken place.

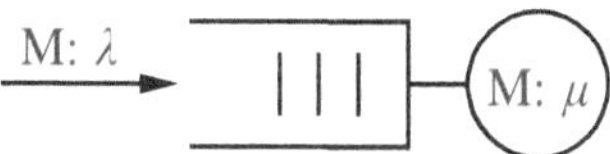


**Figure 5.3:** M/M/1 model

The used model is shown in figure 5.3. The Markovian arrival process has the arrival rate $\lambda$, the Markovian service process has the service rate $\mu$, and the buffer size is infinite.

Relevant values in this domain of simulation speed-up, are the expected error over the value space of the investigated random variable and the resulting number of samples needed to achieve the target error limit.

**Simulation setup** The analytical calculations and the simulations presented in this chapter have been conducted with the following default parameters: The arrival rate $\lambda$ is normalised to be 1 per time unit. Thus, the load $\eta$ can be varied by changing the the service rate $\mu$. The simulation time is $t_s = 10^7$ time units.

### 5.3.1 Dynamic simulation method

First, the dynamic simulation method is analysed to calculate the exact values of all properties of interest and to use these values as reference to compare the dynamic simulation method with the short-term dynamic simulation method with respect to correlation, relative error and, finally, the resulting gain which can be achieved regarding the simulation speed-up.

#### 5.3.1.1 Markov chain of M/M/1 model

The intention is to investigate the simulation method according to the correlation and error properties, see section 2.4 on page 16. For the formulas which are used in that section and in [Sch87], as well as in this chapter, a Markov chain is needed in which the transition probabilities between the states are available instead of transition rates. This is why a Discrete Time Markov Chain (DTMC) is needed.

The considered M/M/1 model, however, is a continuous time model in which the considered events, namely arrivals and departures of jobs, happen at arbitrary points of time on an continuous time axis. One way to build a DTMC for the model is to create an embedded Markov Chain in which the states represent the arrival resp. departure occupancy. Every time an arrival resp. a departure occurs, the occupancy of the system is evaluated, adding the arrived job afterwards resp. subtracting the departed job before. The arrivals resp. departures act as a clock for the discrete points of time at which the state transitions take place. For the M/M/1 model this possibility would indeed be a choice, since the state probabilities would be the same as in the corresponding Continuous Time Markov Chain (CTMC). This is because the memorylessness of the Markov process makes it irrelevant for the state probabilities at which point of time the occupancy is evaluated.

For the simulation, this possibility would be easy to setup. For the calculations conducted in this chapter, however, the model gets more complicated, since many more state transitions are possible. The calculation of stationary state probabilities, transition probabilities, local correlation and relative error gets less clear.

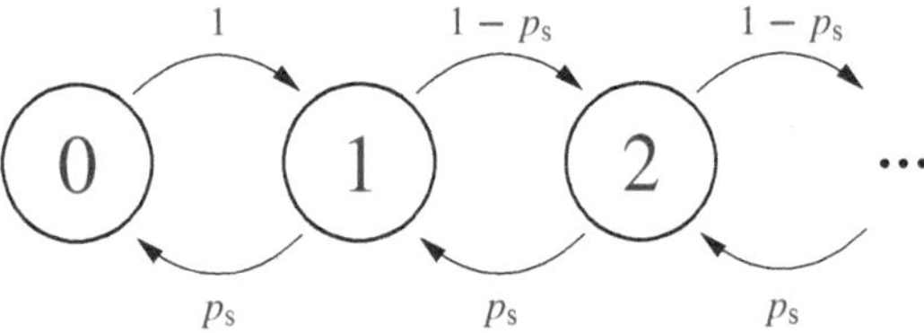


**Figure 5.4:** Discrete time Markov chain of M/M/1 model

Thus, in the chosen model the system occupancy (including queue and server), described by the discrete random variable $X$, is evaluated at every change, i. e., at arrivals as well as at departures. This is possible because both the arrival process as well as the service process are Markovian. The state transitions take place at the discrete points of time of the process resulting from the aggregation of the arrival and the service process. With this approach, the model is represented by a birth-death process, and the corresponding Markov chain is shown in figure 5.4.

A Markov chain which has even more properties equivalent to those of the CTMC can be created by adding a loop transition at state 0, see figure 5.5. The loop is indicated by the transition probability $p_{s0}$, which is zero in figure 5.4. To achieve an exact match with the state probabilities of the CTMC, it has been analytically verified (appendix A) to be

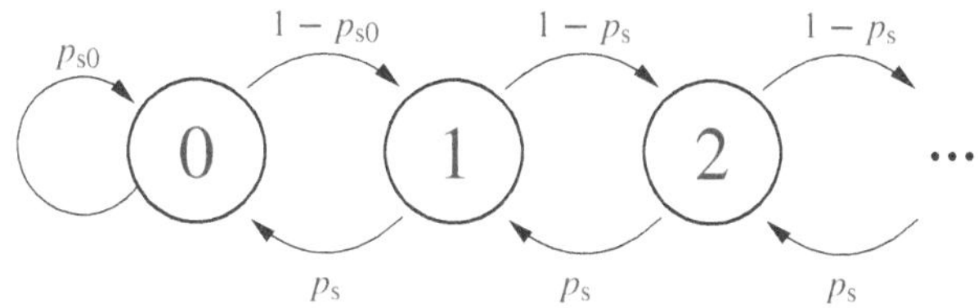


**Figure 5.5:** DTMC of M/M/1 model with extra loop transition at state 0

$$p_{s0} = p_s \tag{5.1}$$

The local correlation, which will be discussed later, differs between the Markov chains in figure 5.4 and figure 5.5 only for the first two states.

From simulation perspective, the loop transition is not intuitive since it does not correspond to a real system event like an arrival or a departure of a job. This means, the simulation would introduce an artificial aspect. This model, however, is chosen because the representation of the system is better regarding the state probabilities making some case differentiations for state 0 unnecessary.

The special behaviour that has to be included into the simulation procedure to represent the loop transition, is to generate an event similar to a departure event when the system is empty (state 0). The scheduled time for this event has to be generated with the same random number generator used for the real departure events. It can be considered as drawing the service time of a virtual job in state 0 causing a state transition back to state 0 after the service time has run down. This special event, however, has to be treated as a conditional event. A state change from a higher state (>0) to the next lower state means that the next departure event has been scheduled to occur before the next arrival event, leading to the transition probabilities $p_s$ resp. $1 - p_s$. The extra transition at state 0 can only be modelled correctly if the special event is deleted as soon as a new arrival occurs. This is because the special event does not describe the departure of a real job, but it only models the condition that the departure of the virtual job takes place ahead of the next arrival. If the special event is not deleted, the the DTMC model would correspond to a CTMC model as in figure 5.6.

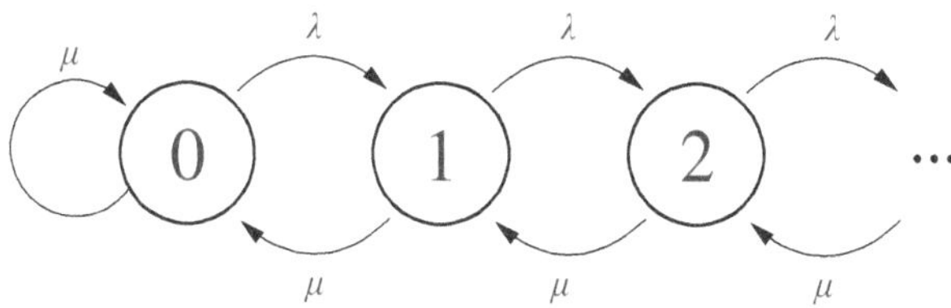


**Figure 5.6:** CTMC of M/M/1 model with extra loop transition at state 0

Since the state probabilities for a CTMC as in figure 5.6 are time weighted, the loop transition does not change these probabilities. For the DTMC case, each loop transition increases the counter for entering state 0, and thus, increases the state probability, since in the DTMC case the probabilities are based on relative frequencies of occurrences. Not deleting the special event would lead to a higher steady state probability for state 0 than it is for the DTMC in figure 5.5.

#### 5.3.1.2 Transition and state probabilities

The probability for the event that the next service completion (departure) precedes the next arrival, is described by $p_s$. The transition probabilities $p_{ij}$[1] derived from figure 5.5 are

$$\begin{aligned} p_{ij} &= \begin{cases} P\{\text{next departure before next arrival}\} & \text{for } i = j+1,\ j = 0\ldots\infty \\ P\{\text{next arrival before next departure}\} & \text{for } i = j-1,\ j = 1\ldots\infty \\ 0 & \text{otherwise} \end{cases} \\ &= \begin{cases} p_s & \text{for } i = j+1,\ j = 0\ldots\infty \\ 1-p_s & \text{for } i = j-1,\ j = 1\ldots\infty \\ p_s & \text{for } i = 0,\ j = 0 \\ 0 & \text{otherwise} \end{cases} \end{aligned} \tag{5.2}$$

A transition probability matrix $\mathbf{p}$ obeys the rule

$$\underline{P}(x) \cdot \mathbf{p} = \underline{P}(x) \tag{5.3}$$

with $\underline{P}(x)$ being the column vector of the steady state probabilities. Now, the transition probability matrix $\mathbf{p}$ can be generated as follows

$$\mathbf{p} = \begin{pmatrix} p_s & 1-p_s & 0 & \cdots & & \\ p_s & 0 & 1-p_s & 0 & \cdots & \\ 0 & p_s & 0 & 1-p_s & 0 & \cdots \\ 0 & 0 & p_s & 0 & \ddots & \ddots \end{pmatrix} \tag{5.4}$$

Given the arrival rate $\lambda$ and the service rate $\mu$ and with this the utilisation $\eta = \lambda/\mu$, $p_s$ can be calculated as shown in equation (5.5). The random variables $A$ resp. $D$ describe the time until the next arrival resp. departure event will happen.

[1] In this document, the intuitive form $p_{ij} = P(j|i)$ is used, describing a transition from state $i$ to state $j$.

$$
\begin{aligned}
p_A(t) &= \lambda e^{-\lambda t}, \qquad p_D(t) = \mu e^{-\mu t} \\
p_s &= P(A > D) \\
&= \int_0^{\infty} p_A(t) \int_0^t p_D(\tau) d\tau dt \\
&= \frac{\mu}{\lambda + \mu} = \frac{1}{1 + \eta}
\end{aligned} \tag{5.5}
$$

The Markov property of the arrival and the service process make $p_A(t)$ and $p_D(t)$ independent of whether the last event was an arrival or a departure and how much time has elapsed since then.

The stationary state probabilities can be calculated with a system of linear equations:

$$
P(0) = P(0) \cdot p_s + P(1) \cdot p_s \tag{5.6}
$$

$$
P(x) = P(x-1) \cdot (1 - p_s) + P(x+1) \cdot p_s \quad \text{for } x = 1 \ldots \infty \tag{5.7}
$$

$$
\sum_x P(x) = 1 \tag{5.8}
$$

From equations (5.6) and (5.7), $P(x)$ can be derived by induction:

$$
P(x) = P(x-1) \cdot \frac{1 - p_s}{p_s} \qquad \text{for } x \geq 1 \tag{5.9}
$$

From this, together with equation (5.8), it follows

$$
P(x) = \frac{2p_s - 1}{p_s} \cdot \left( \frac{1 - p_s}{p_s} \right)^x \tag{5.10}
$$

Using equation (5.5) for $p_s$, $P(x)$ and $E[X]$ can be expressed in terms of $\eta$, as shown in the following.

$$
P(x) = (1 - \eta) \cdot \eta^x \tag{5.11}
$$

$$
E[X] = \sum_{x=0}^{\infty} x P(x) = \frac{\eta}{1 - \eta} \tag{5.12}
$$

Figure 5.7 shows the probability mass functions for two example loads $\eta$ using equations (5.10) and (5.11). Although the probability mass function is only valid for discrete – in this case integer – values of $x$, the points in the figure are connected by a line, for better readability.

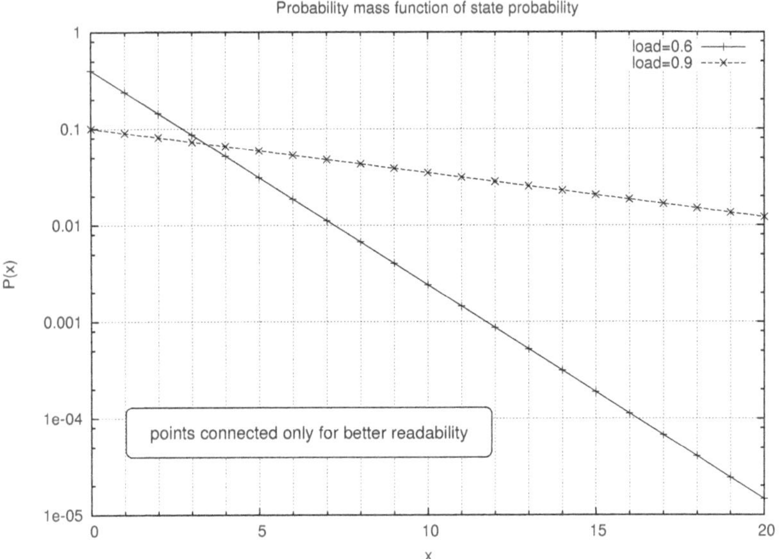


**Figure 5.7:** Probability mass function of the state probability

#### 5.3.1.3 Local correlation

To evaluate the benefit the STD simulation technique has on the error development and with this on the simulation run time, the local correlation has to be considered. It is the main factor of the relative error in the evaluation method LRE (Limited Relative Error), see section 2.4. The 2-node Markov chain, see [Sch87], with its transition probabilities is shown in figure 5.8.

The transition probabilities $p_0(x)$ and $p_1(x)$ of the 2-node Markov chain of the considered model are calculated as shown in equations (5.13) and (5.14). These are the original formulas from [Sch87] and section 2.4 with the adaptations that the model considered there has a finite state space representing a random variable with a continuous value space.

$$p_0(x) = \frac{1}{1-G_x} \cdot \sum_{r=0}^{x} \left( P(r) \sum_{s=x+1}^{\infty} p_{rs} \right) \tag{5.13}$$

$$p_1(x) = \frac{1}{G_x} \cdot \sum_{r=x+1}^{\infty} \left( P(r) \sum_{s=0}^{x} p_{rs} \right) \tag{5.14}$$

These original formulas for $p_0(x)$ and $p_1(x)$ include all possible transitions from all states to all other states. The equations (5.17) and (5.18) for the used model are much simpler since only transitions to neighbouring states are possible in the birth-death Markov chain of the model:

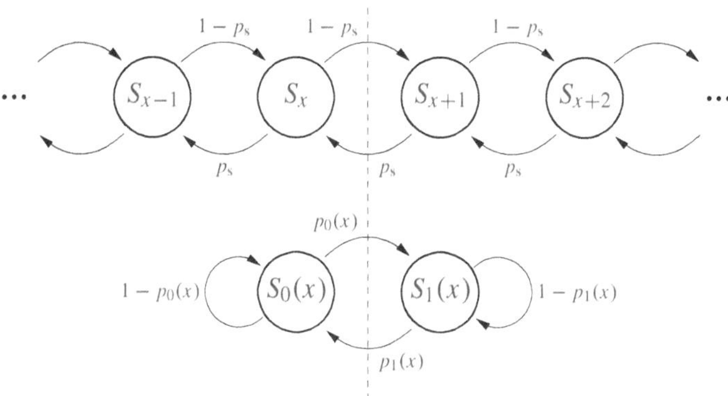


**Figure 5.8:** 2-node Markov chain for M/M/1 model

$$p_{x,x+1} = P(x) \cdot (1-p_s) = P(x) \cdot \frac{\eta}{1+\eta} \tag{5.15}$$

$$p_{x+1,x} = P(x+1) \cdot p_s = P(x+1) \cdot \frac{1}{1+\eta} \tag{5.16}$$

$$p_0(x) = \frac{1}{1-G_x} \cdot p_{x,x+1} \tag{5.17}$$

$$p_1(x) = \frac{1}{G_x} \cdot p_{x+1,x} \tag{5.18}$$

The parameter $G_x$ represents the cumulative state probability of all the states to the *right* of $x$, i. e., the CCDF, while $1-G_x$ corresponds to the states left of $x$ including $x$, i. e., the CDF.

$$G_x = P(X > x) = \sum_{r=x+1}^{\infty} P(r), \qquad 1-G_x = \sum_{r=0}^{x} P(r) \tag{5.19}$$

$$G_x = G(x) = \eta^{x+1} \tag{5.20}$$

Simplifying $p_0(x)$ and $p_1(x)$ with the help of equations (5.10) and equations (5.17) and (5.18), leads to

$$\begin{aligned}
p_0(x) &= \frac{1}{1-G_x}\cdot P(x)\cdot\frac{\eta}{1+\eta} &(5.21)\\
&= \frac{1}{1-(1-\eta)\cdot\sum_{r=x+1}^{\infty}\eta^r}\cdot(1-\eta)\eta^x\cdot\frac{\eta}{1+\eta}\\
&= \frac{1-\eta}{1+\eta}\cdot\frac{\eta^{x+1}}{1-(1-\eta)\cdot\eta^{x+1}\cdot\sum_{r=0}^{\infty}\eta^r}\\
&= \frac{1-\eta}{1+\eta}\cdot\frac{\eta^{x+1}}{1-(1-\eta)\cdot\eta^{x+1}\cdot\frac{1}{1-\eta}}\\
&= \frac{1-\eta}{1+\eta}\cdot\frac{\eta^{x+1}}{1-\eta^{x+1}}\\
p_1(x) &= \frac{1}{G_x}\cdot P(x+1)\cdot\frac{1}{1+\eta} &(5.22)\\
&= \frac{1}{(1-\eta)\cdot\sum_{r=x+1}^{\infty}\eta^r}\cdot(1-\eta)\eta^{x+1}\cdot\frac{1}{1+\eta}\\
&= \frac{1-\eta}{1+\eta}\cdot\frac{\eta^{x+1}}{(1-\eta)\cdot\eta^{x+1}\cdot\sum_{r=0}^{\infty}\eta^r}\\
&= \frac{1-\eta}{1+\eta}\cdot\frac{\eta^{x+1}}{(1-\eta)\cdot\eta^{x+1}\cdot\frac{1}{1-\eta}}\\
&= \frac{1-\eta}{1+\eta}\cdot\frac{\eta^{x+1}}{\eta^{x+1}}\\
&= \frac{1-\eta}{1+\eta}.
\end{aligned}$$

Finally, the local coefficient of correlation $\rho(x)$ is calculated as

$$\rho(x) = 1-(p_0(x)+p_1(x)) \tag{5.23}$$

Using equations (5.21) and (5.22), equation (5.23) leads to

$$\rho(x) = 1-\frac{1-\eta}{1+\eta}\cdot\frac{1}{1-\eta^{x+1}} \tag{5.24}$$

Figure 5.9 shows the local correlation for different loads. For the comparison between simulation results and analytical calculation, the values of the simulation are connected by a line while the analytical values are represented by points. For higher $x$ (the state number representing the number of jobs in the system), the simulation results show increased variations resulting from higher relative error for higher $x$. Therefore, the relative error is also shown in a separate plot below, for which the same line types have been used as the corresponding curves for the correlation. For lower loads, the higher $x$-values have a low probability. This is why the relative error

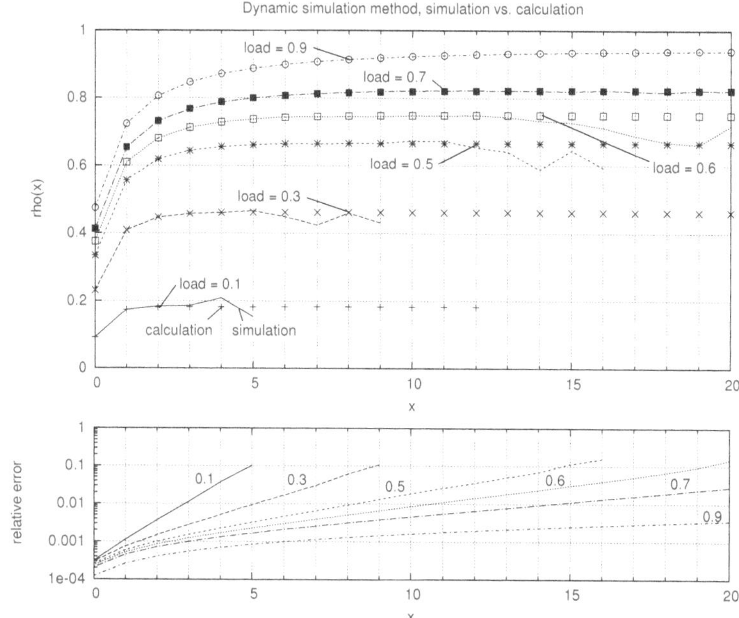


Simulation vs. analytical calculation, $10^7$ time units simulated

**Figure 5.9:** Local correlation vs. state, dynamic simulation method

for the simulation results for the lower loads is higher than for the higher load curves, resulting in shorter curves for the correlation values. For this, see section 5.3.1.4.

Figure 5.10 shows the local correlation from another perspective. For different states it shows the correlation vs. the load. Here, only the analytical calculation is shown, since for higher $x$ and lower load the relative error for the simulation results will be very high. This would lead to a high deviation of the simulated values from the analytical curves, making the plot unclear. In figure 5.9 it is demonstrated how well the analytical equations and the simulations match.

Figure 5.11 shows the development of the two contributing parts of the correlation coefficient $\rho(x)$, namely $p_0(x)$ shown in figure 5.11 a) vs. state $x$ and load $\eta$. In figure 5.11 b), $p_0(x)$ is additionally shown as an array of curves for some example $x$-values. Further, the figure shows $p_1(x)$ (the darker curve) only vs. the load $\eta$ since it is independent of $x$, see equation (5.22). For high load, the transition probability to the right $p_0(x)$ remains low. On the other side, the probability for a transition to the left $p_1(x)$ starts at 1 and converges to 0 for high load. This together with equation (5.23) leads to a correlation rising to 1 for high loads and increasing $x$.

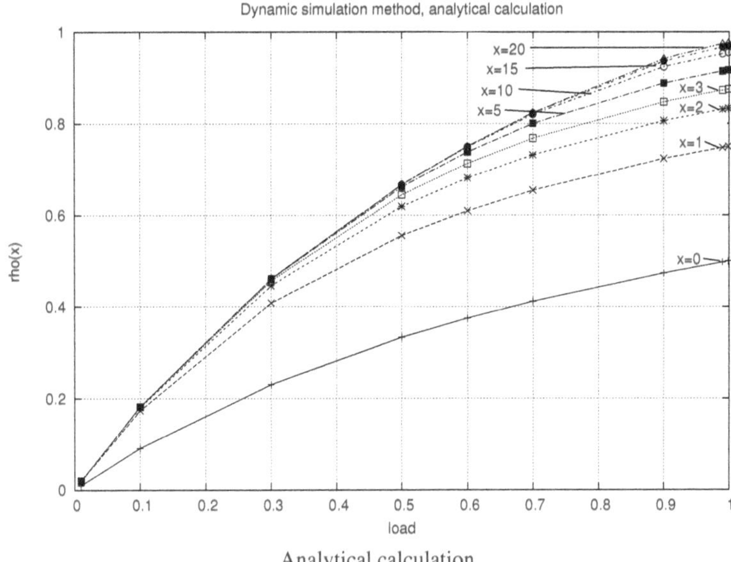


Analytical calculation

**Figure 5.10:** Local correlation vs. load, dynamic simulation method

#### 5.3.1.4 Relative error

As mentioned in equation (5.19), for the discrete random variable used for this model, $G_x$ is defined as $G_x = P(X > x) = G(x)$. Compared to the description of the local correlation in section 2.4 where a continuous state space is considered, the $x$ used here is mapped to $i-1$ of section 2.4: $x \leftrightarrow i-1$. The cumulative frequency $\upsilon_x$ is

$$\upsilon_x = \sum_{r=x+1}^{\infty} h_r = \tilde{G}_x \cdot n_s \tag{5.25}$$

Here, $\tilde{G}_x$ is the obtained cumulative relative frequency of the states *right* of $x$, $h_x$ the obtained relative frequency of state $x$, and $n_s$ the absolute number of evaluated samples of the random variable $X$.

Using $\tilde{\rho}(x)$ for the simulated local correlation, the relative error $d_G(x)$ of a simulated sample is according to equation (2.9) as follows

$$d_G(x) = \left( \frac{1}{n_s} \cdot \frac{1-\tilde{G}_x}{\tilde{G}_x} \cdot \frac{1+\tilde{\rho}(x)}{1-\tilde{\rho}(x)} \right)^{1/2} \tag{5.26}$$

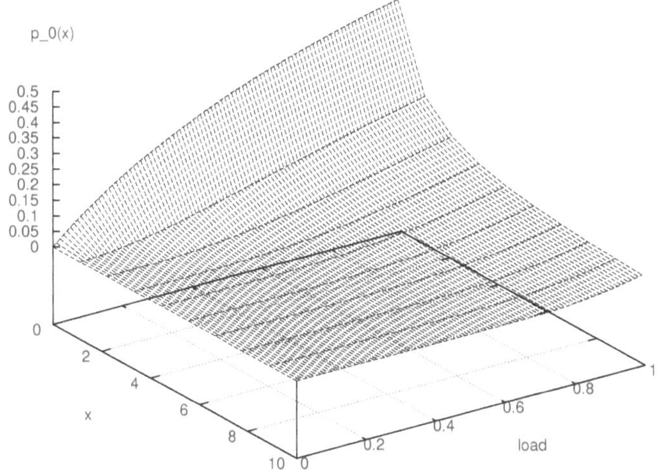


a) $p_0(x)$ vs. load and x

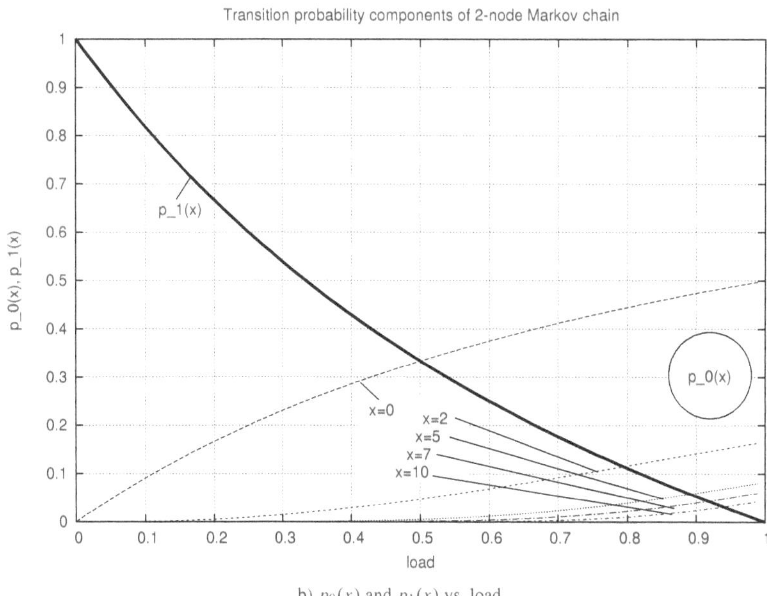


b) $p_0(x)$ and $p_1(x)$ vs. load

**Figure 5.11:** Transition probability components of the 2-node Markov chain

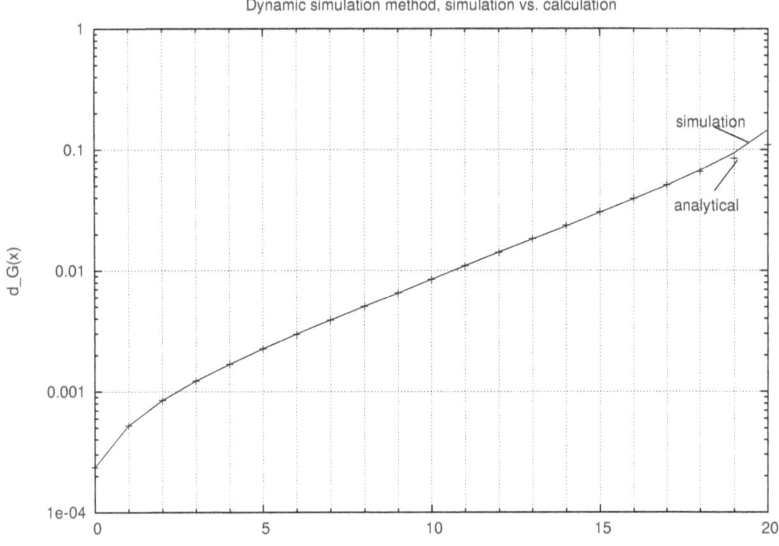


Simulation vs. analytical calculation, $10^7$ time units simulated, $\eta = 0.6$

**Figure 5.12:** Relative error, dynamic simulation method

In this analytical investigation, the theoretical relative error $d_G^*(x)$ is needed to compare the different simulation approaches. For this, the theoretical $G_x$ is used instead of the simulated $\tilde{G}_x$, as well as $\rho(x)$ instead of $\tilde{\rho}(x)$:

$$d_G^*(x) = \left(\frac{1}{n_s} \cdot \frac{1 - G_x}{G_x} \cdot \frac{1 + \rho(x)}{1 - \rho(x)}\right)^{1/2} \tag{5.27}$$

Figure 5.12 shows the relative error for a load of $\eta = 0.6$, comparing simulation and analytical results. The simulation results are shown by the solid line while the curve for the analytical results shows the expected relative error from equation (5.27) and is represented by points.

### 5.3.2 Short-term dynamic simulation

While in the dynamic simulation each state transition corresponds to an event in the model, in short-term dynamic simulation extra state transitions are introduced. These transitions take place between STD windows and are responsible for the fact that the Markov chain is not a simple birth-death model any longer. The reason is the independence of the starting state of the STD window from the states before. An STD transition, which can also be called extra transition, starts from the last state of an STD window and cannot only go to a neighbouring state but to any state of the state space. The STD transition probability from state $i$ to the target state $x$ is equal to the steady state probability $P(x)$ of the state $x$, independently of state $i$.

#### 5.3.2.1 Markov chain of M/M/1 model

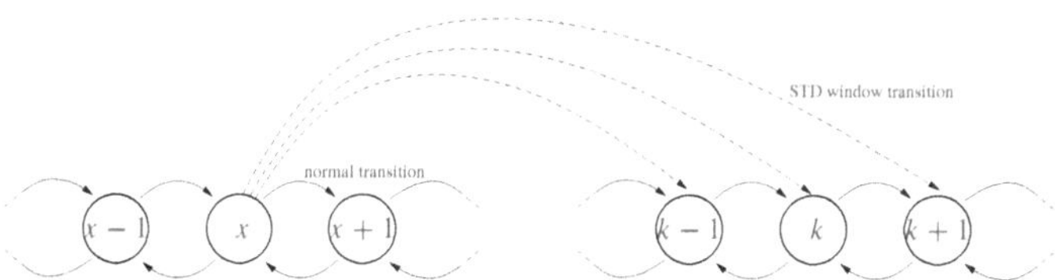


**Figure 5.13:** Markov chain of M/M/1 model with STD extensions

Figure 5.13 shows the Markov chain of the model, similar to the one in figure 5.5, but with the extra transitions for new STD windows.

The probability of an extra transition is $p_{\text{ex}}$. It represents the number of new STD windows relative to the total number of state transitions. Its calculation will be discussed in section 5.3.2.2 in detail.

These extra transitions cause the calculation of the transition probabilities to be extended by another term. Compared to the transition probability $p_{ij,\text{dyn}}$ of the dynamic simulation, the $p_{ij,\text{STD}}$ of the short-term dynamic simulation is as follows:

$$p_{ij,\text{STD}} = (1 - p_{\text{ex}}) \cdot p_{ij,\text{dyn}} + p_{\text{ex}} \cdot p_{ij,\text{ex}} \tag{5.28}$$

While in the dynamic simulation the transition probabilities to non-neighbouring states are zero, see equation (5.2), this is different for the extra transition:

$$\begin{aligned} p_{ij,\text{ex}} &= P\{\text{last state previous window = i|next starting state = j}\} \\ &= P\{\text{next starting state = j}\} \\ &= P(j) \end{aligned} \tag{5.29}$$

From equation (5.28), it now follows:

$$p_{ij,\text{STD}} = (1 - p_{\text{ex}}) \cdot p_{ij,\text{dyn}} + p_{\text{ex}} \cdot P(j) \tag{5.30}$$

#### 5.3.2.2 Extra transition probability

The mean duration of an STD window is represented by $\overline{T}_{\mathrm{w}}$. Keeping the total simulation time $t_{\mathrm{s}}$ constant, the mean number of STD windows $\overline{N}_{\mathrm{w}} = t_{\mathrm{s}}/\overline{T}_{\mathrm{w}}$. In case of constant STD window lengths in units of time, $\overline{T}_{\mathrm{w}}$ resp. $\overline{N}_{\mathrm{w}}$ are simply $t_{\mathrm{w}}$ resp. $n_{\mathrm{w}}$. Considering now constant STD windows, the times of the state changes follow a plain Poisson process with the intensity parameter $\lambda + \mu$ with a mean number of state transitions per STD window of $\lambda + \mu$.

The mean number of evaluated states during an STD window $\overline{n}_{\mathrm{s,w}}$ is equal to the number of state transitions within the window plus one. The additional one is because the starting state and the last state of an STD window are evaluated. It is now

$$\overline{n}_{\mathrm{s,w}} = (\lambda + \mu) \cdot \overline{T}_{\mathrm{w}} + 1 \tag{5.31}$$

The mean total number of evaluated states within the simulation $\overline{n}_{\mathrm{s}}$, i. e., the mean number of values obtained within the simulation, is

$$\overline{n}_{\mathrm{s}} = \overline{n}_{\mathrm{s,w}} \cdot \overline{N}_{\mathrm{w}} \tag{5.32}$$

Finally, $p_{\mathrm{ex}}$ is

$$p_{\mathrm{ex}} = \frac{1}{\overline{n}_{\mathrm{s,w}}} \cdot \frac{\overline{N}_{\mathrm{w}} - 1}{\overline{N}_{\mathrm{w}}} = \frac{1}{(\lambda + \mu) \cdot \overline{T}_{\mathrm{w}} + 1} \cdot \frac{t_{\mathrm{s}} - \overline{T}_{\mathrm{w}}}{t_{\mathrm{s}}} \tag{5.33}$$

The last STD window has no transition to another one, which is why the number of extra transitions is one below the number of STD windows $\overline{N}_{\mathrm{w}}$, inducing the term $\frac{\overline{N}_{\mathrm{w}}-1}{\overline{N}_{\mathrm{w}}}$.

Usually, the number of STD windows should not be too small, since otherwise the difference from the pure dynamic simulation is small, which in turn applies also for the gain that can be achieved by the STD simulation. Thus, assuming $\overline{N}_{\mathrm{w}} \gg 1$ or equivalently $\overline{T}_{\mathrm{w}} \ll t_{\mathrm{s}}$, the following approximation can be used:

$$p_{\mathrm{ex}} \approx \frac{1}{\overline{n}_{\mathrm{s,w}}} = \frac{1}{(\lambda + \mu) \cdot \overline{T}_{\mathrm{w}} + 1} \tag{5.34}$$

#### 5.3.2.3 Local correlation

Knowing the transition probability of an extra transition, the transition probabilities of the 2-node Markov chain according to figure 5.8 and equations (5.17) and (5.18) can be adapted for the STD case. Using equations (5.15) and (5.16), we have

$$p_0(x) = \frac{1}{1-G_x} \cdot \left( (1-p_{\text{ex}})\, p_{x,x+1} + p_{\text{ex}} \cdot \left( \sum_{r=0}^{x} \left( P(r) \sum_{s=x+1}^{\infty} P(s) \right) \right) \right) \tag{5.35}$$

$$p_1(x) = \frac{1}{G_x} \cdot \left( (1-p_{\text{ex}}) p_{x+1,x} + p_{\text{ex}} \cdot \left( \sum_{r=x+1}^{\infty} \left( P(r) \sum_{s=0}^{x} P(s) \right) \right) \right) \tag{5.36}$$

The sums in the extra part of $p_0(x)$ and $p_1(x)$ can be simplified:

$$\begin{aligned}
&\sum_{s=x+1}^{\infty} P(s) = G_x, \qquad \sum_{s=0}^{x} P(s) = 1 - G_x \\
&\sum_{r=0}^{x} (P(r) \cdot G_x) = G_x \cdot \sum_{r=0}^{x} P(r) = G_x\,(1-G_x) \\
&\sum_{r=x+1}^{\infty} (P(r) \cdot (1-G_x)) = (1-G_x) \cdot \sum_{r=x+1}^{\infty} P(r) = (1-G_x)\, G_x
\end{aligned} \tag{5.37}$$

These simplifications lead with equations (5.35) and (5.36) to

$$\begin{aligned}
p_0(x) &= \frac{1}{1-G_x} \cdot \left((1-p_{\text{ex}})p_{x,x+1} + p_{\text{ex}} \cdot (1-G_x)G_x\right) \\
&= (1-p_{\text{ex}}) \cdot \frac{1}{1-G_x} p_{x,x+1} + p_{\text{ex}} \cdot G_x
\end{aligned} \tag{5.38}$$

$$\begin{aligned}
p_1(x) &= \frac{1}{G_x} \cdot \left((1-p_{\text{ex}})p_{x+1,x} + p_{\text{ex}} \cdot G_x(1-G_x)\right) \\
&= (1-p_{\text{ex}}) \cdot \frac{1}{G_x} p_{x+1,x} + p_{\text{ex}} \cdot (1-G_x)
\end{aligned} \tag{5.39}$$

Finally, the local correlation for the STD case, $\rho_{\text{STD}}(x)$, can be calculated:

$$\begin{aligned}
\rho_{\text{STD}}(x) &= 1 - \left( (1-p_{\text{ex}}) \cdot \left( \frac{1}{1-G_x} \cdot p_{x,x+1} + \frac{1}{G_x} \cdot p_{x+1,x} \right) + p_{\text{ex}} \cdot (G_x + 1 - G_x) \right) \\
&= 1 - \left( (1-p_{\text{ex}}) \cdot \left( \frac{1}{1-G_x} \cdot p_{x,x+1} + \frac{1}{G_x} \cdot p_{x+1,x} \right) + p_{\text{ex}} \right) \\
&= (1-p_{\text{ex}}) + \left( (1-p_{\text{ex}}) \cdot \left( \frac{1}{1-G_x} \cdot p_{x,x+1} + \frac{1}{G_x} \cdot p_{x+1,x} \right) \right) \\
&= (1-p_{\text{ex}}) \cdot \left( 1 - \left( \frac{1}{1-G_x} \cdot p_{x,x+1} + \frac{1}{G_x} \cdot p_{x+1,x} \right) \right) \\
&= (1-p_{\text{ex}}) \cdot \rho(x)
\end{aligned} \tag{5.40}$$

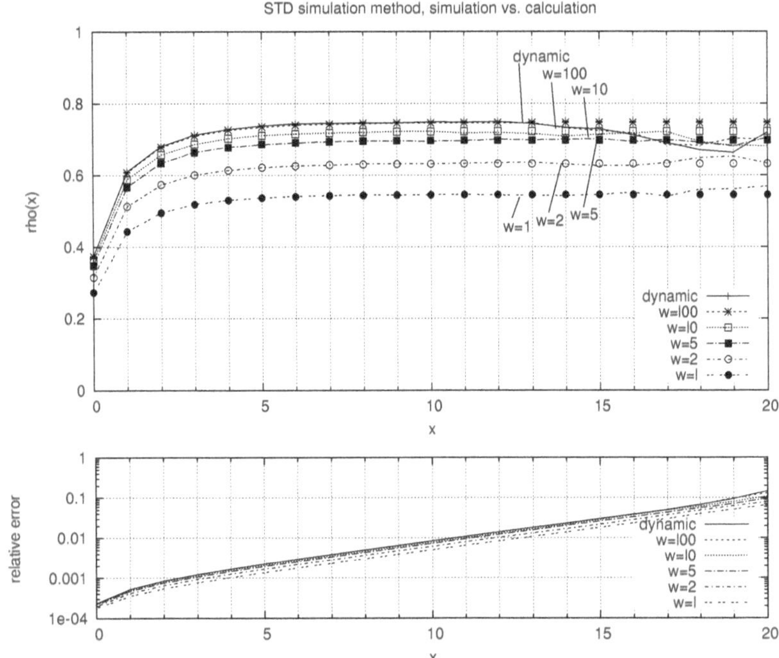


Simulation vs. analytical calculation, $10^7$ time units simulated, $\eta = 0.6$
STD window sizes from 1 to 100 time units
Maximum relative error at $x = 20$ between 0.06 and 0.2

**Figure 5.14:** Local correlation, STD simulation method

The local correlation $\rho_{\text{STD}}(x)$ in the STD case deviates from the dynamic case simply by the factor $1 - p_{\text{ex}}$.

Figure 5.14 shows the local correlation for the STD simulation method with different STD window sizes. They are indicated by "$w =$" with the subsequent number in the legend of the figure and shown by the different dashed lines, while the solid line shows for comparison the correlation for a dynamic simulation. The lines show the simulation results while the points show the values of the analytical calculations. Further, the relative error for the simulation results is shown to explain the deviation from the calculated values especially for larger $x$. This results from the lower probability for larger $x$. Shorter STD windows lead to lower relative error.

For higher values of $x$, the relative error increases because compared to the relative error for the lower $x$-values, the $G(x)$ is lower. Combining the STD simulation concept with the RESTART technique for rare event simulation can be a way to speed-up simulations which are evaluating such rare events with the STD simulation concept. This is discussed in section 5.4.2.

The value for $w$ represents the STD window size in time units. Smaller STD windows result in lower local correlation. Using the default parameters mentioned on page 61, the curve with "$w = 1$" corresponds to a simulation with an STD window size of $\overline{T}_{\mathrm{w}} = 1$ time units and a mean number of $(\lambda + \mu) \cdot \overline{T}_{\mathrm{w}} = 8/3$ transitions per STD window, resulting in a mean number of $\overline{n}_{\mathrm{s,w}} = 11/3$ evaluated states (for the given $\eta = 0.6$), see equation (5.31).

#### 5.3.2.4 Relative error

In the STD case, the expected relative error is calculated similarly to the relative error of the dynamic case, since the local correlation is the only variable in equation (5.27) which is different in the STD case, provided the number of samples $n$ is kept constant. The relative error $d^{*}_{G}(x)$ in the STD case is now

$$\begin{aligned} d^{*}_{G}(x) &= \left( \frac{1}{n_{\mathrm{s}}} \cdot \frac{1 - G_x}{G_x} \cdot \frac{1 + \rho_{\mathrm{STD}}(x)}{1 - \rho_{\mathrm{STD}}(x)} \right)^{1/2} \qquad (5.41) \\ &= \left( \frac{1}{n_{\mathrm{s}}} \cdot \frac{1 - G_x}{G_x} \cdot \frac{1 + \rho(x)(1 - p_{\mathrm{ex}})}{1 - \rho(x)(1 - p_{\mathrm{ex}})} \right)^{1/2} \end{aligned}$$

Figure 5.15 show the relative error for STD simulations with different STD windows sizes, including the dynamic simulation for comparison. Smaller STD windows lead to lower relative error.

#### 5.3.2.5 Relative deviation

To confirm the fact that a smaller relative error caused by a smaller local correlation indeed leads to more accurate results, figure 5.16 shows the distribution function of the model used above, and further its accuracy. Figure 5.16 a) on the left side shows the complementary cumulative distribution function which corresponds to the probability mass function for the load $\eta = 0.6$ in figure 5.7. Increasing STD window sizes and further the dynamic simulation show decreasing accuracy. To emphasise this fact, figure 5.16 b) on the right side shows the relative deviation from the calculated distribution function for the different STD window sizes.

### 5.3.3 Group correlation

Another way to evaluate the STD simulation technique with respect to correlation is to consider groups (sequences) of states and calculate the local correlation of the sequence of these groups. A basis for the correlation can either be the starting state of a group or the mean value of the states of a group.

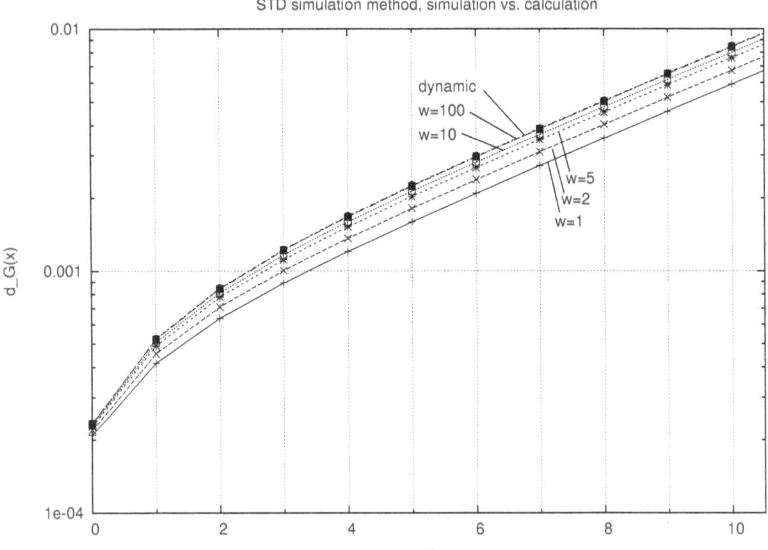


Simulation vs. analytical calculation, $10^7$ time units simulated, $\eta = 0.6$

**Figure 5.15:** Relative error, STD simulation

In the following, $P_j^{(h)}(x)$ describes the probability that the $j^{\text{th}}$ state in the $h^{\text{th}}$ group is $x$, thus, $P_0^{(h)}(x)$ corresponds to the starting state in group $h$. The index $n$ is defined to correspond to the last state of the group, assuming in the first place that the number of states per group is constant.

### 5.3.3.1 Multi-step transition probabilities

In addition to the transition probability matrix of equation (5.4), probability vectors are needed to describe multi-step transition probabilities. The probability vector for the $k^{\text{th}}$ state in group $h$ is $\mathbf{v}_k^{(h)}$. It is reached from the starting vector $\mathbf{v}_0^{(h)}$ by $k$-fold multiplication with the transition probability matrix $\mathbf{p}$:

$$\mathbf{v}_k^{(h)} = \mathbf{v}_0^{(h)} \cdot \mathbf{p}^k \tag{5.42}$$

In dynamic simulation, the last state of a group is the first state of the next group, leading to $\mathbf{v}_0^{(h)} = \mathbf{v}_n^{(h-1)}$ for $h \geq 1$. The steady state probability vector is simply called $\mathbf{v}$.

The single $n$-step transition probability $p_{ij}(n)$ from $i$ to $j$ is

$$p_{ij}(n) = \left(\mathbf{p}^n\right)_{i+1,j+1} \tag{5.43}$$

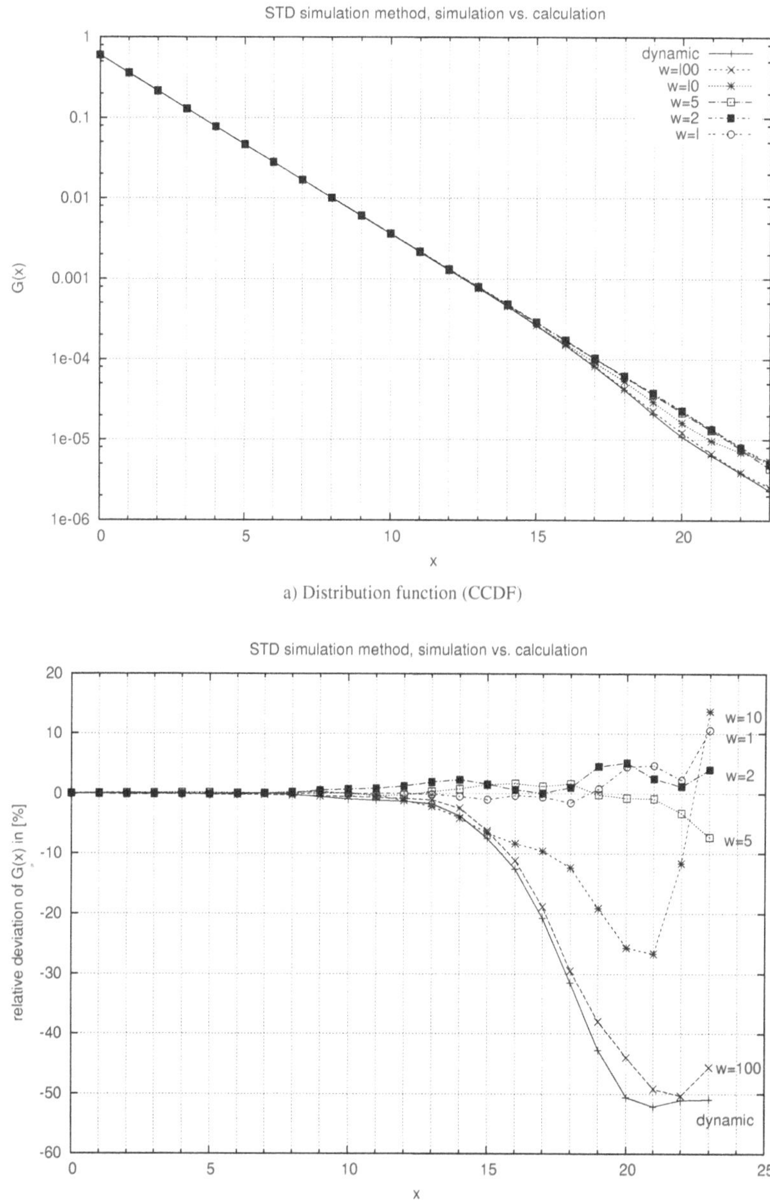


b) Relative deviation

STD simulation, $10^7$ time units simulated

**Figure 5.16:** CCDF and relative deviation from analytical CCDF

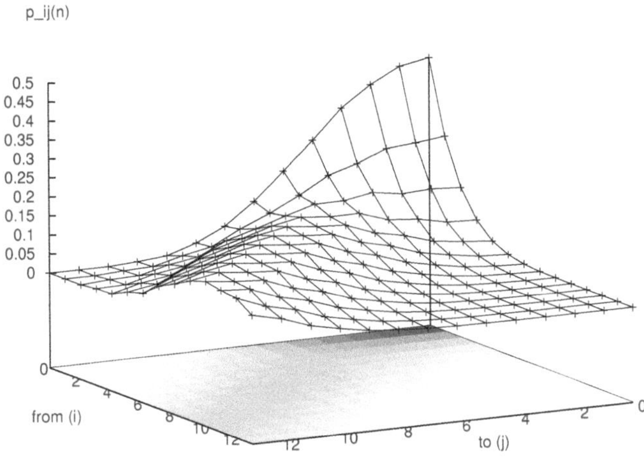


$n = 10$ steps, $\eta = 0.6$, superposition of $n$-step and $(n+1)$-step matrices

**Figure 5.17:** $n$-step transition probability matrix

Since the states start at 0 and the row and column indices in matrices always start at 1, the transition probability $p_{ij}(n)$ is the element $(i+1, j+1)$ of the above matrix. This mapping of indices and state numbers applies generally to the matrices in the following sections, as well as to the single-step transition probability matrix $\mathbf{p}$.

The birth-death property of the used model results in the fact that from a certain state another certain state can only be reached in either an odd or an even number of steps depending on the distance of the states. E. g., state 5 can be reached from state 3 in two steps or in four, but not in three or in five. An exception which breaks this rule is the transition at state 0 to state 0. Therefore, state 5 can be reached from state 3 also in 9, 10, 11 or more steps.

This birth-death property causes the diagonals of $\mathbf{p}$ in equation (5.4) to be zero beyond state 0. A $k$-step transition is represented by $\mathbf{p}^k$. For an odd $k$, the main diagonal is (almost) zero as in the single-step matrix $\mathbf{p}$. For an even $k$, the main diagonal is non-zero but the neighbouring diagonals are (almost) zero. The number of non-zero diagonals is $k+1$, surrounded by an upper and a lower triangle with zero values. Due to the self-transition at state 0, the upper left part of the $k$-step transition matrix is non-zero. This applies to the matrix elements of which the sum of the column index and the row index is not larger than $k+1$. This is why the diagonals are referred to as *(almost) zero*.

An example multi-step transition probability matrix is shown graphically in figure 5.17. The (almost) zero diagonals and most of the non-zero values are surrounded by zero values, prohibits to connect neighbouring values in the plot by lines because this would make the plot very unclear.

It is possible to superpose the subsequent matrices $\mathbf{p}^k$ and $\mathbf{p}^{k+1}$. After dividing the elements by 2, the resulting matrix is again a stochastic matrix and can be interpreted as follows: With the probability 0.5 each, either $k$ steps or $k+1$ steps will be performed. Based on this, the plotted matrix represents the probability to reach the target state from the starting state in either $k$ or $k+1$ steps. The figure only shows the first part for $i \leq 12$ and $j \leq 12$ of the matrix which has infinite dimensions.

#### 5.3.3.2 Multi-step local correlation

Considering now the Markov chain applying the $n$-step transitions, the birth-death property is lost. Therefore, the 2-node Markov chain needed to calculate the local correlation for this case cannot be calculated as easily as in the original model. The corresponding $n$-step transition probabilities $p_0(x,n)$ and $p_1(x,n)$, equations (5.44) and (5.45), are less simple than the ones of equations (5.17) and (5.18).

$$p_0(x,n) = \frac{1}{1-G_x} \cdot \sum_{r=0}^{x} \left( P(r) \sum_{s=x+1}^{\infty} p_{rs}(n) \right) \tag{5.44}$$

$$p_1(x,n) = \frac{1}{G_x} \cdot \sum_{r=x+1}^{\infty} \left( P(r) \sum_{s=0}^{x} p_{rs}(n) \right) \tag{5.45}$$

These equations can be simplified by using matrix expressions to make them more clear. Further, this can be useful for working with mathematical software to calculate these values. First, a special identity matrix is introduced which is used to zero out rows or columns of vectors and matrices in multiplication. Equation (5.46) shows this $\mathbf{I}_x$. In a real identity matrix, simply all $a_i$ parameters are 1. For the form used in the lower part of the equation, the dimension of the upper left sub-matrix is $x$.

$$\mathbf{I}_x = \begin{pmatrix} a_1 & 0 & \cdots & & \\ 0 & a_2 & 0 & \cdots & \\ \vdots & 0 & a_3 & 0 & \\ & \vdots & 0 & \ddots & \end{pmatrix} \quad \text{with } a_i = \begin{cases} 1 & \text{for } 1 \leq i \leq x \\ 0 & \text{otherwise} \end{cases} \tag{5.46}$$

$$\mathbf{I}_x = \left( \begin{array}{c|c} I & 0 \\ \hline 0 & 0 \end{array} \right) \qquad \mathbf{I} - \mathbf{I}_x = \left( \begin{array}{c|c} 0 & 0 \\ \hline 0 & I \end{array} \right)$$

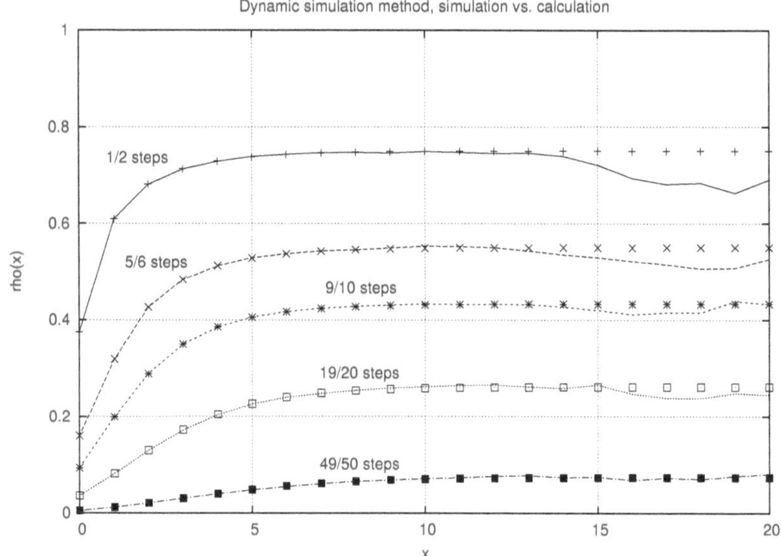


Simulation vs. analytical calculation, $10^7$ time units simulated, $\eta = 0.6$
Maximum relative error at $x = 20$ between 0.06 and 0.14

**Figure 5.18:** Multi-step local correlation

Using this, equations (5.44) and (5.45) can be written as follows:

$$\begin{aligned}\mathbf{w}(x) &= \frac{1}{1-G_x}\cdot(\mathbf{vI}_{x+1})\cdot(\mathbf{p}\,(\mathbf{I}-\mathbf{I}_{x+1}))\\ p_0(x) &= \sum_i w_i = \mathbf{w}(x)\cdot(1\ 1\ \ldots)^{\mathsf{T}}\end{aligned} \tag{5.47}$$

$$\begin{aligned}\mathbf{u}(x) &= \frac{1}{G_x}\cdot(\mathbf{v}\,(\mathbf{I}-\mathbf{I}_{x+1}))\cdot(\mathbf{pI}_{x+1})\\ p_1(x) &= \sum_i u_i = \mathbf{u}(x)\cdot(1\ 1\ \ldots)^{\mathsf{T}}\end{aligned} \tag{5.48}$$

The matrix $\mathbf{I}_x$ zeroes out the appropriate columns right of $x$ in the probability vector $\mathbf{v}$ and the transition probability matrix $\mathbf{p}$. The complementary matrix $\mathbf{I}-\mathbf{I}_x$ does the same with the columns left of $x$ including $x$. To be precise: $x+1$ is used as the index since the row and column indices of matrices always start with 1, but in this case $x$ represents the state, and $x$ is in the range $0\ldots\infty$.

Figure 5.18 shows the multi-step local correlation corresponding to the single-step local correlation in figure 5.14. Different step sizes are shown, in pairs of two, meaning as stated above, that the figure shows the superposition of two adjacent matrices regarding the number of steps. The simulations have been performed with the same sample size.

#### 5.3.3.3 Complete STD window

Groups can simply match complete STD windows. In this case, the group correlation will be zero in STD simulation and the starting probability vector of any group is independent of the last probability vector of any other group. Instead, it is equal to the steady state probability vector: $\mathbf{v}_0^{(h)} = \mathbf{v}$.

The reduction of the correlation compared to the dynamic simulation is therefore equal to the multi-step correlation value corresponding to the window size in figure 5.18. In accordance with the single-step local correlation, it decreases with increasing STD window size.

#### 5.3.3.4 Mean value of groups

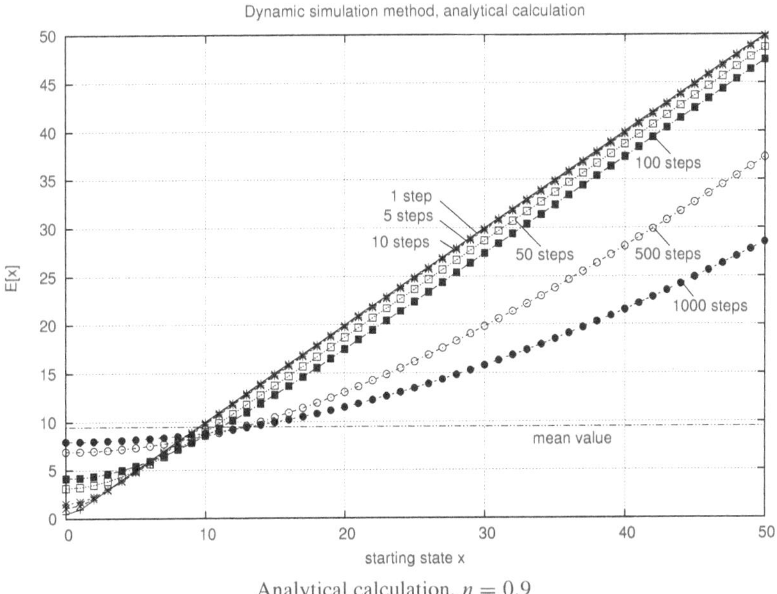


Analytical calculation, $\eta = 0.9$

**Figure 5.19:** Mean value of groups of different size, depending on starting state $x_0$

Now, groups of fixed size are considered, i. e., a group consists of a fixed number of samples $n_{s,g}$. The mean value of a group, i. e., the expected value for the state during the simulation of these $n_{s,g}$ samples, depends on the starting state. A mean probability vector $\overline{\mathbf{v}}^{(h)}$ can be calculated by adding all the $k$-step probability vectors $\mathbf{v}_k^{(h)}$ of a group $h$ and dividing it by the number of the vectors:

$$\overline{\mathbf{v}}^{(h)} = \frac{1}{n_{s,g}} \sum_{k=0}^{n_{s,g}-1} \mathbf{v}_k^{(h)} = \frac{1}{n_{s,g}} \sum_{k=0}^{n_{s,g}-1} \mathbf{v}_0^{(h)} \mathbf{p}^k = \frac{1}{n_{s,g}} \mathbf{v}_0^{(h)} \sum_{k=0}^{n_{s,g}-1} \mathbf{p}^k \tag{5.49}$$

An mean transition probability matrix $\overline{\mathbf{p}}(n_{s,g})$ for $(n_{s,g} - 1)$ steps ($n_{s,g}$ samples) can be derived from this. It represents the relative frequency of the states within a group of size $n_{s,g}$ depending on the starting state. Each row $\overline{\mathbf{p}}_i(n_{s,g})$, with $i = 1,2,..,$ of the matrix represents the corresponding starting state $i - 1$. The matrix is

$$\overline{\mathbf{p}}(n_{s,g}) = \frac{1}{n_{s,g}} \sum_{k=0}^{n_{s,g}-1} \mathbf{p}^k \tag{5.50}$$

With this, equation (5.49) can be written as equation (5.51). The starting state probability vector $\mathbf{v}_0^{(h)}$ of group $h$ multiplied with the relative frequency matrix results in the relative frequency of the states in group $h$.

$$\overline{\mathbf{v}}^{(h)} = \mathbf{v}_0^{(h)} \cdot \overline{\mathbf{p}}(n_{s,g}) \tag{5.51}$$

The starting state vector $\mathbf{v}_0^{(h)}$ of group $h$ in a dynamic simulation results from the multiplication of the first starting state vector $\mathbf{v}_0^{(1)}$ and the transition probability matrix $\mathbf{p}$ to the power of $(n_{s,g} - 1) \times (h - 1)$. Assuming, without loss of generality, that a dynamic simulation always starts in state 0, $\mathbf{v}_0^{(h)}$ is equal to the first row of the matrix $\mathbf{p}^{(n_{s,g}-1)\cdot(h-1)}$.

The relative frequency vector of the starting states $\mathbf{v}_0$ depends on the number of simulated groups $N_g$ which converges against $\mathbf{v}$ for large values of $N_g$ and $\mathbf{v}_0$.

$$\mathbf{v}_0 = \frac{1}{N_g} \sum_{h=1}^{N_g} \mathbf{v}_0^{(h)} = \frac{1}{N_g} \sum_{h=1}^{N_g} \left( \mathbf{v}_0^{(1)} \left( \mathbf{p}^{n_{s,g}-1} \right)^{h-1} \right) = \frac{1}{N_g} \mathbf{v}_0^{(1)} \sum_{h=1}^{N_g} \mathbf{p}^{(n_{s,g}-1)\cdot(h-1)} \tag{5.52}$$

The column vector $\mathbf{x}(n_{s,g})$ of the mean state values depending on the starting state is calculated by multiplying the mean transition probability matrix $\overline{\mathbf{p}}(n_{s,g})$ by a column vector representing the state numbers:

$$\mathbf{x}(n_{s,g}) = \overline{\mathbf{p}}(n_{s,g}) \cdot (0\ 1\ 2\ \ldots)^{\mathsf{T}} \tag{5.53}$$

Figure 5.19 shows $\mathbf{x}(n_{s,g})$ for different values of $n_{s,g}$ vs. $x$. The steady state expectation value for the state – indicated by the horizontal line which is named *mean value* in the graph – is not approached or more slowly for starting states far away from the expectation value and for smaller group sizes.

The mean state $\overline{x}^{(h)}$ of group $h$ can be simply calculated by multiplying the mean state probability vector $\overline{\mathbf{v}}^{(h)}$ by a column vector representing the state numbers, or by multiplying the starting state vector $\mathbf{v}_0^{(h)}$ of group $h$ with the mean state vector $\mathbf{x}(n_{\mathrm{s,g}})$:

$$\overline{x}^{(h)} = \overline{\mathbf{v}}^{(h)} \cdot (0\ 1\ 2\ \ldots)^{\mathsf{T}} = \mathbf{v}_0^{(h)} \cdot \mathbf{x}(n_{\mathrm{s,g}}) \tag{5.54}$$

A total mean state value $\overline{x}$ is simply as in equation (5.55). For large $n_{\mathrm{s}}$, i. e., for large $n_{\mathrm{s,g}}$ and/or large $N_{\mathrm{g}}$, it converges against the expected value $E[X]$, equation (5.12).

$$\overline{x} = \mathbf{v}_0 \cdot \mathbf{x}(n_{\mathrm{s,g}}) \tag{5.55}$$

#### 5.3.3.5 Distribution of mean value of groups

With the calculations above, the mean value for a group depending on the starting state is obtained. To find the correlation among the mean values, however, the distribution of the mean values in a group has to be calculated.

All possible paths of the length of the group and starting from a certain state $x_0$, have to be identified and classified by their weight, i. e., by the sum of the states visited within the path. Further, each path has to be weighted by its probability, which is simply the product of the probabilities of all involved transitions.

The frequency of these sums provides the distribution, and dividing the sums by the length of the path (the size of the group), the mean value distribution is obtained. These are all conditional for starting at state $x_0$.

Keeping the matrix operations, the mean values are represented by the sums to have them represented by the rows and columns of the corresponding matrices. The matrix showing the probabilities of the group sums $s$ for a given starting state $x_0$ is $\mathbf{P}(s|x_0)$. The rows of $\mathbf{P}(s|x_0)$ represent the starting states, the columns represent the sums.

Next, the relative frequency of the sums independent of the starting state has to be calculated. In equation (5.56), the relative frequency of the starting states $\mathbf{v}_0$, taken from equation (5.52), is multiplied by the conditional probability matrix $\mathbf{P}(s|x_0)$. $S$ is defined as the random variable for the state sums, and $\mathbf{s}$ is the row vector containing the relative frequencies $P_S(i)$ for $i = 0, 1, \ldots$

$$\mathbf{s} = (P_S(0)\ P_S(1)\ \ldots) = \mathbf{v}_0 \cdot \mathbf{P}(s|x_0) \tag{5.56}$$

Figure 5.20 shows for different group sizes $n_{\mathrm{s,g}}$ the probability mass function according to the probabilities represented by the vector $\mathbf{s}$. The solid lines show calculation results from the above equations, the points on them show the values obtained by simulation. Further, the relative error for the simulation results is shown.

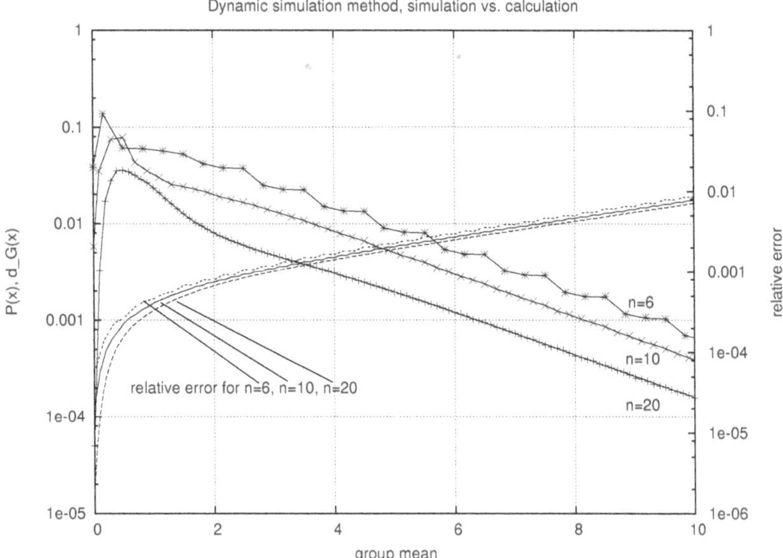


Simulation vs. analytical calculation, $1.3 \cdot 10^7$ groups simulated, $\eta = 0.6$

**Figure 5.20:** Probability mass function for group-mean for different group sizes

#### 5.3.3.6 Correlation of mean value of groups

To calculate the correlation between the sums – and by that between the mean values – the transition probabilities between the sums are needed. A direct relation between the sums does not exist. Instead, the conditional probability matrix $\mathbf{P}(s|x_0)$ of a sum $s$ for a given starting state $x_0$ is known. Further, the conditional probability matrix $\mathbf{P}(x_n|s)$ of the starting state $x_n$ of the next group for a sum $s$ of the current group can be calculated.

In the same procedure in which the matrix $\mathbf{P}(s|x_0)$ is built, the parts of $\mathbf{P}(x_n|s)$ are calculated simultaneously. For a certain starting state $x_0$ all possible paths are calculated, and for the sum of each path the two possible successor-states are figured out. These two states are the possible starting states of the next group. The counting of these states is weighted with the product of the path probability and the transition probability to the state. By this way, for each $x_0$ a matrix $\mathbf{P}(x_n|s,x_0)$ for the $x_n$ with the two conditions $s$ and $x_0$ is calculated. Multiplying the matrix with the probability of $x_0$ and summing up the results leads to the required $\mathbf{P}(x_n|s)$:

$$\mathbf{P}(x_n|s) = \sum_{x_0} P(x_0) \cdot \mathbf{P}(x_n|s,x_0) \tag{5.57}$$

To finally calculate the local correlation for the group mean values according to equations (5.13) and (5.14), the transition probability matrix for the sums $\mathbf{p}_g$ is

$$\mathbf{p}_g = \mathbf{P}(x_n|s) \cdot \mathbf{P}(s|x_0) \tag{5.58}$$

Similarly to equations (5.44) and (5.45), equations (5.59) and (5.60) calculate the two components needed for the local correlation. Here, $s$ is used instead of $x$ to express the fact that these considerations are about the group sums of the states and not about the states themselves.

$$p_0(s) = \frac{1}{1-G_s} \cdot \sum_{i=0}^{s} \left( P_S(i) \sum_{j=s+1}^{\infty} p_{ij} \right) \tag{5.59}$$

$$p_1(s) = \frac{1}{G_s} \cdot \sum_{i=s+1}^{\infty} \left( P_S(i) \sum_{j=0}^{s} p_{ij} \right) \tag{5.60}$$

In matrix notation, similar to equations (5.47) and (5.48), they get the following form:

$$p_0(s) = \frac{1}{1-G_s} \cdot (\mathbf{s} \cdot \mathbf{I}_{s+1}) \cdot \left(\mathbf{p}_g \cdot (1 \ldots 1\, 0 \ldots)^{\mathsf{T}}\right) \tag{5.61}$$

$$p_1(s) = \frac{1}{G_s} \cdot (\mathbf{s}\,(\mathbf{I} - \mathbf{I}_{s+1})) \cdot \left(\mathbf{p}_g \cdot (0 \ldots 0\, 1 \ldots)^{\mathsf{T}}\right) \tag{5.62}$$

Just like equation (5.23), $\rho(s)$ applies to the group means:

$$\rho(s) = 1 - (p_0(s) + p_1(s)) \tag{5.63}$$

Figure 5.21 shows similar to the multi-step local correlation of figure 5.18 the local correlation of the group mean values. The sample sizes of these simulations are the same for all group sizes.

#### 5.3.3.7 Relative error

All the formulas leading to the group correlation are for the dynamic simulation. In the STD simulation, the group correlation $\rho(s)$ is zero per definition for the case that the group size is the same as the STD window size. Thus, the values calculated for the group mean indicate the gain $\gamma(\rho_{\max})$ with respect to the reduction of the correlation, and in turn the reduction of the relative error of the group mean values.

Equation (5.64) shows the relative error for the group mean values for a $\rho(s)$ as calculated in equation (5.63) and with $\rho(s) = 0$ for the STD case.

$$d_G^*(s) = \left( \frac{1}{N_g} \cdot \frac{1-G_s}{G_s} \cdot \frac{1+\rho(s)}{1-\rho(s)} \right)^{1/2} \tag{5.64}$$

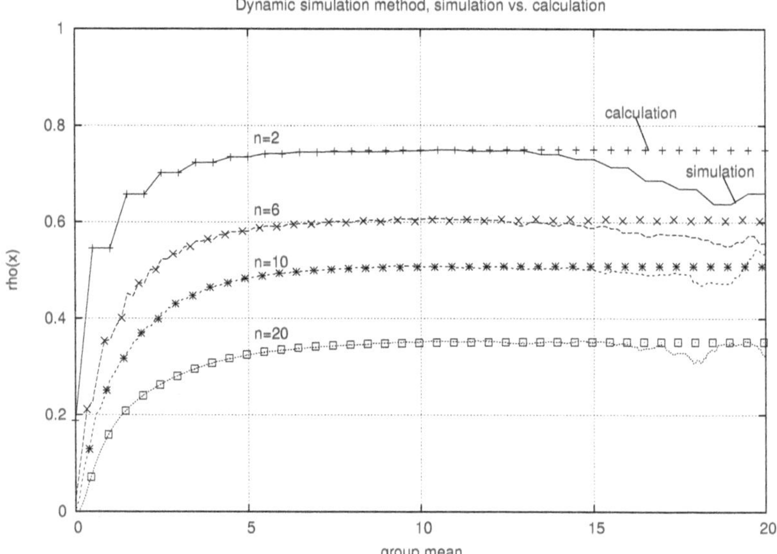


Simulation vs. analytical calculation, $1.3 \cdot 10^7$ groups simulated, $\eta = 0.6$
Maximum relative error at $x = 20$ between 0.1 and 0.15

**Figure 5.21:** Group-mean local correlation

### 5.3.4 Gain

As stated in chapter 2, the increased statistical accuracy of the STD simulation technique can be expressed by the decreased simulation run-time resp. by the decreased sample size needed to reach a predefined simulation status. This status can be represented by the relative error for the investigated value space.

#### 5.3.4.1 STD simulation

Fixing $d$ in equations (5.27) and (5.41) leads to the gain in simulation time $\gamma\,(p_{\text{ex}},\rho_{\max})$.

$$\gamma\,(p_{\text{ex}},\rho_{\max}) = \frac{n_{\text{s,dyn}}}{n_{\text{s}}} = \frac{\frac{1+\rho_{\max}}{1-\rho_{\max}}}{\frac{1+\rho_{\max}(1-p_{\text{ex}})}{1-\rho_{\max}(1-p_{\text{ex}})}} = \frac{\rho_{\max}^2 - \rho_{\max}\frac{p_{\text{ex}}}{1-p_{\text{ex}}} - \frac{1}{1-p_{\text{ex}}}}{\rho_{\max}^2 + \rho_{\max}\frac{p_{\text{ex}}}{1-p_{\text{ex}}} - \frac{1}{1-p_{\text{ex}}}} \tag{5.65}$$

Here, $\rho_{\max}$ means the $\rho(x)$ which causes the maximum relative error within the evaluated value space for x. In the given model, the relative error is monotone in $x$, thus, the $\rho_{\max} = \rho(x_{\max})$.

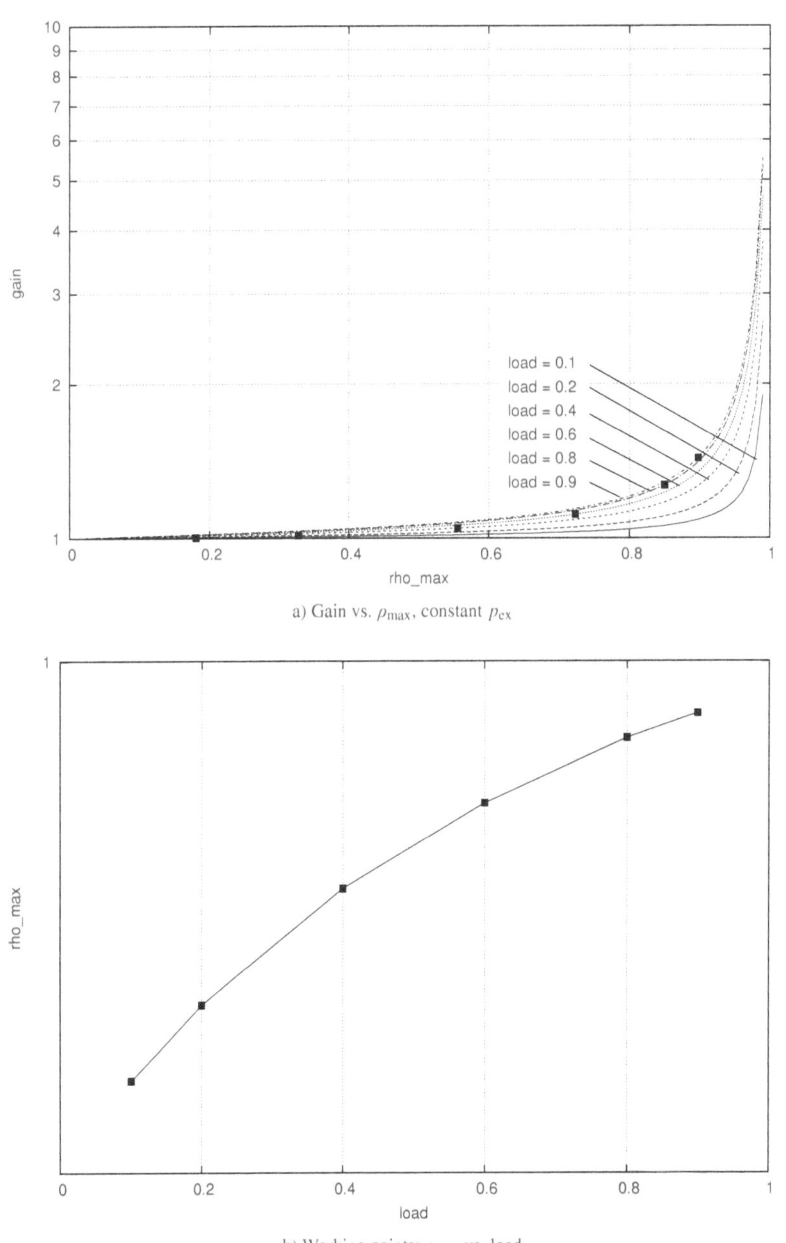


a) Gain vs. $\rho_{\max}$, constant $p_{ex}$

b) Working points: $\rho_{\max}$ vs. load

**Figure 5.22:** Gain $\gamma\,(p_{ex},\rho_{\max})$ vs. $\rho_{\max}$ (STD simulation), different loads $\eta$

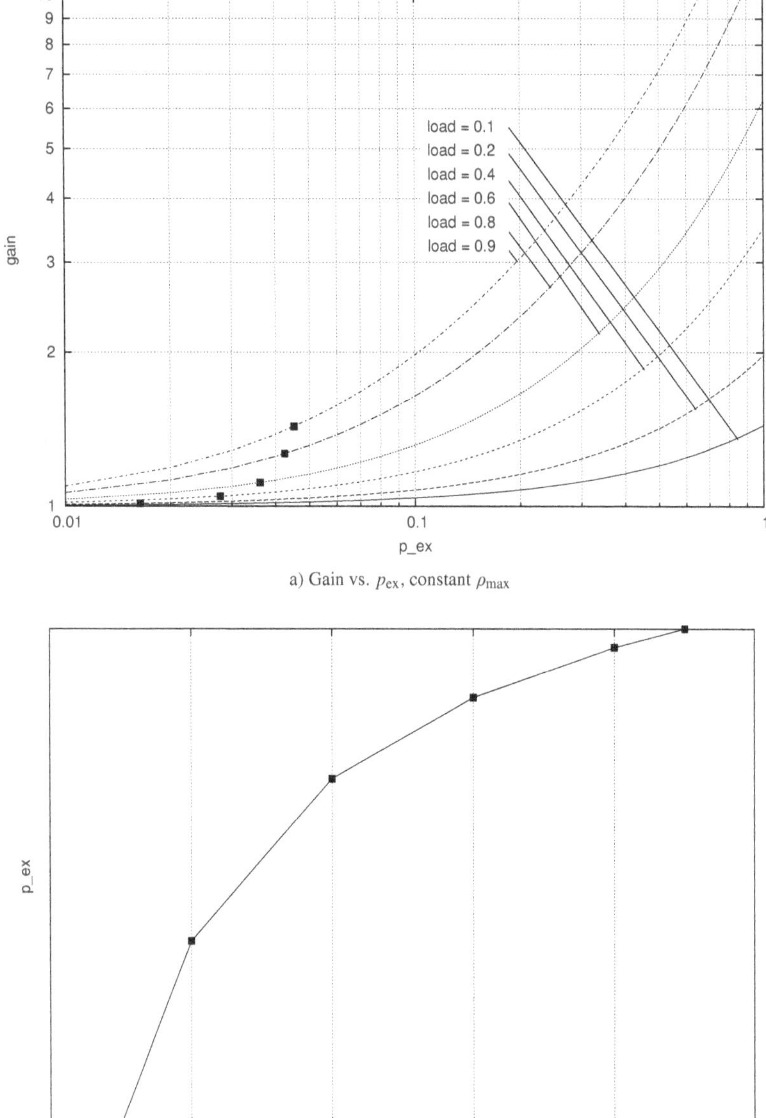


a) Gain vs. $p_{ex}$, constant $\rho_{max}$



b) Working points: $p_{ex}$ vs. load

**Figure 5.23:** Gain $\gamma(p_{ex}, \rho_{max})$ vs. $p_{ex}$ (STD simulation), different loads $\eta$

Figures 5.22 and 5.23 show the achieved gain of equation (5.65) regarding the number of samples required to fall below the specified maximum relative error. In figure 5.22 a), $\rho_{max}$ is changed within its limits (from 0 to 1) while the $p_{ex}$ is kept constant by keeping the STD window size constant, $t_w = 10$ in this example. In figure 5.23 a), $p_{ex}$ is changed from 0 to 1 while $\rho_{max}$ is kept constant. For these plots, $x = 20$ has been used for $\rho_{max}$. The plots at the bottom (figures 5.22 b) and 5.23 b)) show the working points for the different loads $\eta$ in the plots above. These working points are the values of the parameter that has been kept constant for the plots, e. g., for a load of $\eta = 0.6$, a constant $p_{ex} \approx 3.6 \cdot 10^{-2}$, figure 5.23 b), is used for the corresponding plot in figure 5.22 a), and a constant $\rho_{max} \approx 0.723$, figure 5.22 b), is used for the corresponding plot in figure 5.23 a).

The constant $\rho_{max}$ corresponds to the model with a load of $\eta = 0.6$ at the position $x = 20$ which makes a $\rho_{max} \approx 0.723$. The single point in each figure shows the gain value for the constant parameter of the other figure.

#### 5.3.4.2 Group correlation

The group correlation can be considered as a kind of smoothed correlation compared to the first order local correlation investigated in sections 5.3.1 and 5.3.2. The group correlation with respect to the group mean values has been investigated in section 5.3.3 to show that the statistical accuracy of an STD simulation is higher compared to a pure dynamic simulation even if the system is considered from a different perspective.

To confirm this, the gain based on the relative error of the group mean values, see section 5.3.3.7, has to be calculated. Equation (5.66) shows this gain $\gamma(\rho_{max})$ regarding the required number of groups $N_g$ calculated the same way as in equation (5.65) by fixing the relative error $d_G^{\bullet}(s)$ and using a $\rho_{max}$ of the considered range of $s$:

$$\gamma(\rho_{max}) = \frac{1+\rho_{max}}{1-\rho_{max}} > 1 \qquad \forall \quad \rho_{max} > 0 \tag{5.66}$$

This confirms the above statement, since there is an actual gain $\gamma(\rho_{max}) > 1$.

It is, however, possible to make use of the group correlation as a means for speeding-up the simulation. For this, several aspects have to be taken into account. If only the mean values of the groups are considered relevant, the STD simulation is fast in any case since the required number of groups $N_g$ is always lower than for the corresponding dynamic simulation, independent of the group size $n_{s,g}$. Only for large $n_{s,g}$, the comparison group correlation converges to zero, and the dynamic simulation will be as fast as the STD simulation.

Since the mean value of a group can only be determined after all the associated values have been simulated, the speed-up potential based on the group correlation will not be very large. This is different in situations where the process evaluating the mean values is much more time consuming than the process calculating them. The process evaluating the mean values will be responsible for the decision about reaching the given error limit. This is why it is indeed helpful to have knowledge about the group mean correlation in dynamic and STD simulations. Section 5.4.1 and especially section 5.4.1.2 shows an application.

## 5.4 Combined speed-up approaches

In chapter 4, a combination of the two distinct speed-up approaches RESTART and parallelisation, as introduced in chapter 3 resp. section 2.3, is described already. In this section, additionally, some approaches are discussed that combine the STD simulation concept with each of these.

### 5.4.1 STD and parallelisation

The fact that the STD windows are mutually independent automatically brings up the idea of parallelisation. It appears to be a straightforward approach and it will be basically a load division. It can be compared to a simulation project which starts a set of concurrent simulations with the same model and parametrisation but different random seeds, see section 2.3.2.

#### 5.4.1.1 Scalability

With the usual load division, the set of simulations can be started with a relaxed (a higher) error limit which is a-priori known. Equations (5.26), (5.41) and (5.64) show that the relative error is indirectly proportional to the square root of the sample size. For a total target error level $d_{\text{tot}}$ and a number of $n_{\text{p}}$ processes doing the load division, each process should run with an error limit of $d = \sqrt{n_{\text{p}}} \cdot d_{\text{tot}}$ having each process simulating an equal part of the total sample size. Eventually, this leads to the above overall target error level.

In contrast to that, in STD simulations, the size of the STD windows is a parameter specified before the simulation is started. The number of STD windows needed to reach an error limit can, indeed, be determined during the simulation, see discussion in section 6.5. This means, after the STD window size has been decided according to the evaluation requirements, see discussion in section 6.5.4, all simulations started at the beginning will finish at least one complete STD window. It should be calculated with a task granularity of about the size of an STD window.

The granularity of the tasks into which the simulation can be divided for the concurrent processing, builds a limit for the scalability of the parallelisation approach. If large STD windows are needed, e. g., of 100 s, and 10 processors are available for the simulation, the total simulation time will be at least 1000 s. If a certain target error level is going to be reached after 500 s without knowing this in advance, half of the simulation resources are wasted. For small STD windows, this is less likely to happen.

#### 5.4.1.2 Evaluation

An issue with this kind of parallelisation is the combination of the results. This is similar to the completely independent runs of the usual kind of load division. Either all observations, i. e., the complete obtained sample, is relevant, and in turn all samples have to be stored for some kind of central integration after all simulations have finished. Another possibility is to have the separate simulations runs pre-evaluate the results on basis of single STD windows and have a certain entity merge these statistics afterwards, see the brief discussion in section 3.2.3.

In any case, the results of the different STD windows have to be combined in some way. For that, they have to be communicated to some central entity. While large STD windows have the disadvantage of limited granularity for a decision about the finish of the simulation, simulations with small STD windows need to frequently communicate the (pre-evaluated) results of the STD windows.

It can be more expensive – with respect to resources or time – to communicate results than to produce them. Of course, this depends on the setup. A quite large STD window takes a considerable amount of processing time, and the results will probably not take such long time to be communicated over a network, even if quite detailed information about the complete STD window is transmitted. For smaller STD windows, this is different. The ratio of processing time and frequency of communication of results makes the communication an important part of the total run time.

In cases where only the mean value of certain parameters on the basis of an STD window is relevant, the communication of this result will take less time than the communication of the complete sample or even a pre-evaluated statistics of the parameter. These mean values will not be correlated in the STD simulation, thus, the evaluation process will reach the given error limit more quickly than it would for a dynamic simulation also sending mean values of consecutive sample groups. As a conclusion on this aspect, knowing the mean value correlation in a dynamic simulation helps to justify the application of a parallelised STD simulation for the above mentioned cases with small STD windows and only group mean values being relevant.

### 5.4.2 STD and RESTART

A link between the STD simulation concept and the RESTART method is the starting state for a part of the simulation. The simulation control finishes this part and decides about another system state from which the next part of the simulation is going to be started.

In RESTART, all system states from which the simulation can start a retrial have been collected before by the simulation itself during the elapsed simulation time. Furthermore, all states belonging to the same level have in common the value of the importance function (IF). More precisely, after having obtained a previous value below the threshold, either the current value is exactly the same as the threshold or it has a value above the threshold.

In contrast to this, in the STD simulation concept, a system state building the initial snapshot from which an STD window is started is generated on the basis of the statistical data describing the scenario and its configuration.

#### 5.4.2.1 Specifying initial snapshots

The fact that statistical information is available about the scenario does not make simulation dispensable. In the UMTS application, e. g., see chapter 6, first, the number of users is a random variable, next, the position of each of these users is a set of random variables, and finally, the session and mobility dynamics of each user consisting of many single events is generated on the basis of many random generators. The simulation user is interested in the interactions of the individual simulation objects and the probabilities of certain situations.

An example for the application of this kind of combination is a quite simple queueing network. The evaluation target is the behaviour of a small part of the network under the condition of another small part of the network being in a special state. The components of this small part are included in the importance function (IF) which can simply be the occupancy of a single node, as with the simple model used in section 5.3. The special state is the occupancy having the exact value of a chosen threshold. All states for which the IF is above or equal to the threshold is considered as the target importance region in the following discussion.

For this application, it is assumed that the cumulative probability of the target importance region can be calculated from the statistical information used to generate the snapshots. Possible dependencies of the occupancy of this node from other random parts of the network have to be incorporated into the calculation.

It is now possible to generate the initial snapshots matching the threshold and study the behaviour of the small part of the network model for these situations. This part can be some neighbouring nodes, and the evaluation target is the rare event set and can be the buffer overflows in these nodes. Compared to STD simulations with initial snapshots representing the whole state space, the target can be evaluated much faster.

Since the evaluation target is the total, i. e., unconditional, probability of the evaluation target (the rare event set) $A$, but the simulation provides only the conditional target $A|B$ always reached after starting from a subset (the target importance region) $B$, $P(B)$ is needed. From the statistical information about the state space, the probability $P(B)$ is assumed to be known. If it is possible to reach $A$ not only from $B$, but also from states $\bar{B}$ with the IF below the threshold, i. e., $P(A|\bar{B}) > 0$, however, this has to be respected:

$$P(A) = P(A|B)P(B) + P(A|\bar{B})P(\bar{B}) \tag{5.67}$$

It means, parts of the rare event set are outside the area of the state space which is limited by the threshold. Such configurations should be avoided if possible.

In a more complex UMTS example, see chapter 6, the number of users connected to a certain cell in a scenario is the threshold, and for a quite high number of users, the overload behaviour of the cell is studied. If, e. g., the dropping is counted, according to the statements above, the dropping in cases of less users than the threshold has to be considered. This is possible even in low load situations if users move too far away from the base station and are dropped due to power limitations. From this example it is obvious that the above-mentioned configurations are not always avoidable.

#### 5.4.2.2 History of states

In contrast to RESTART, where a system state has been stored after it has been reached from some other state below the threshold, in STD simulations, system states representing independent initial snapshots of STD windows have to be generated. In principle, these snapshots can be generated in such a way that they match exactly the value of a threshold.

This kind of independent snapshot generation, however, completely lacks any information about the history of the system state represented by the initial snapshot. This is why it cannot be decided whether the generated snapshot can represent a transition from a region below to a region above or equal to the threshold. On the other hand, a transition downwards during the STD window can be detected, and the STD window can be terminated the same way a trace is terminated in RESTART.

One way to cope with this lack of history is to simply ignore this fact. An IF with a single threshold, in the first place, is chosen. Snapshots are generated matching the threshold. STD windows are terminated as soon as the IF falls below the threshold limiting the evaluation to the target importance region.

The usual application of this method combination will be such that the evaluation target is a special quite rare situation. Also the target importance region will not have a very high probability since otherwise the application of this method combination would make no sense. As a consequence, many of the generated initial snapshots will be of no use.

If, e. g., the probability of the target importance region is 10 %, the other 90 % of the generated snapshots are wasted. This is because the IF will fall below the threshold with the very first value following the snapshot. The probability of the target importance region and the effort to generate a snapshot has to be taken into account when planning to apply this approach.

The criteria for terminating an STD window could be chosen differently from the way it would be done in RESTART. In some variations of RESTART, the downward threshold is different from the upward threshold which introduces a hysteresis. Discussions on the efficiency of such variations can be found among others in [VAVA99] and [Gar00]. For the combination of STD and RESTART, however, this hysteresis can at least compensate the lack of history to some extent.

#### 5.4.2.3 Support statistical information

As stated above, usually the total probabilities of events are to be determined. If now the probability of the target importance region is not easy to calculate from the statistical information about the state space, this can be evaluated by an STD simulation.

If the stationary distributions of the states for the initial snapshots and the subsequent dynamic parts of the STD window match very well, see discussion in section 6.5.3, this approach can provide the necessary probability value.

If the probability of the target importance region can be calculated but only with assumptions or other influences on the precision, the method to simulate the probability can at least support the statistical information.

#### 5.4.2.4 Finding initial snapshots

While in section 5.4.2.1 the statistical information about the system states is used to generate specific initial snapshots with no history, such snapshots can also be found by simulation which enables to store these states and provides them with a history. The approach of section 5.4.2.3 is one way to achieve this. The process of evaluating the probability of the target importance region can be utilised for this purpose.

Another possibility is to start from some defined state, e. g., an empty system. In STD simulations, however, long transient phases at the start of the STD windows conflict with the STD window size considerations with respect to efficiency. This will be the case if the mean number of objects in the system is high, e. g., users in the UMTS applications of chapter 6, and furthermore, the arrival rates are small with long dwell times, see Little's law in equation 6.1.

## 5.5 Conclusions

The perspective which is usually taken, is to consider all values for evaluation. For a simple simulation model, it has been proven analytically that the STD simulation method increases the statistical accuracy. The gain that will be achieved by this is shown in section 5.3.4.1.

As expected, the gain is higher for smaller STD windows, since the behaviour of STD simulations become more similar to the behaviour of static simulations with decreasing STD window sizes, and more similar to dynamic simulations with increasing STD window sizes. Depending on the evaluation target, STD window sizes down to quite small ones can still make sense. This is further discussed in chapter 6, especially in sections 6.3 and 6.5.

Some approaches to combine the STD simulation concept with the other simulation speed-up techniques parallelisation and RESTART have been identified and discussed. The efficiency that can be expected from applying these approaches depends on the kind of simulation model and the evaluation targets.

# Chapter 6

# STD Simulation of UMTS Models

After the general concept of STD simulation and its analytical investigation have been presented in chapter 5, this chapter describes the application to UMTS network planning.

## 6.1 Introduction to UMTS

### 6.1.1 Architecture

A UMTS system consists of three main components, namely the UTRAN, the core network (CN), and the user equipment (UE).

As a major difference to 2G systems as GSM, the access technology on the air interface has changed to a CDMA (Code Division Multiple Access) system instead of FDMA (Frequency Division Multiple Access) or TDMA (Time Division Multiple Access). Therefore, another Radio Access Network (RAN) was necessary, the UMTS Terrestrial Radio Access Network (UTRAN). Except from special cases with very small cells and high data-rates where it is reasonable to apply TDD (Time Division Duplexing), in UTRAN FDD (Frequency Division Duplexing) is used.

The main components of the UTRAN are the base stations, also called Node-Bs, and the Radio Network Controllers (RNC). The main functions of a base station which are relevant for this chapter, are transmission and reception on the air interface, physical channel coding, and closed loop power control. The main functions of the RNCs are Radio Resource Control (RRC), admission control, and handover control. The base stations are connected to the core network via the RNCs, and one RNC can support one or more base stations.

As the main function, the core network provides switching, routing and transit for user traffic. The network management functions are also contained in the core network, as are the databases EIR (Equipment Identity Register), HLR (Home Location Register), and VLR (Visitor Location Register).

The user equipment interfaces with the user and the air interface. A UE consists of two components. One is the mobile equipment (ME) which is the radio terminal. The other is the USIM (UMTS Subscriber Identity Module). The latter stores the subscriber information as identity and encryption key, and it performs the authentication. In the following, the user equipment is referred to as a mobile station (MS).

### 6.1.2 Power control

To separate the signals of different mobile stations at the base station, the spreading codes of the connections need to be used since in FDD, the mobile stations transmit at the same frequency in uplink direction. If each MS would transmit at the same power, the received power of a far away MS would be far below the received power of an MS close to the base station because of the higher path loss of the far away MS. The signals of these mobile stations would not be separable anymore.

The closed loop power control algorithm takes care of the fact that the received power per bit of all connections will be equal. Frequent measurements of the received SIR (Signal to Interference Ratio) are performed, and the values are compared to a target SIR. The MS is commanded to either lower or increase it transmission power to reach this target. This measurement cycle is performed at 1500 Hz to prevent any power mismatch among the uplink signals received at the base station.

In contrast to the closed loop power control, the RNC performs an open loop power control. This is to roughly estimate the required transmission power, and it is only used at the beginning of a connection.

### 6.1.3 Regular scenario

Two scenarios representing two different classes are exemplarily shown here. They demonstrate the abilities of the simulation concept, and further, the following sections use variants of these scenarios. This section shows a regular scenario, the next section shows a realistic scenario.

A regular scenario is introduced without any distinction between areas with different surfaces or different heights, and without any other irregularity like buildings or other obstacles. The base stations are arranged in a hexagonal grid because the antennas can form circular-like cells on plain scenarios. For such shapes, the hexagonal arrangement provides the best level of regularity. Circular-like shapes are formed by omni-directional antennas or a little less regular by sectorised antennas with three sectors.

Figure 6.1 shows such a scenario. It consists of seven sites (base stations) with three sectors (cells) each. The sites are indicated by the numbers surrounded by squares, and the antenna sectors are the three short lines originating at the number of each site and identified by another number at the end of the line. As a result, the scenario has 21 cells.

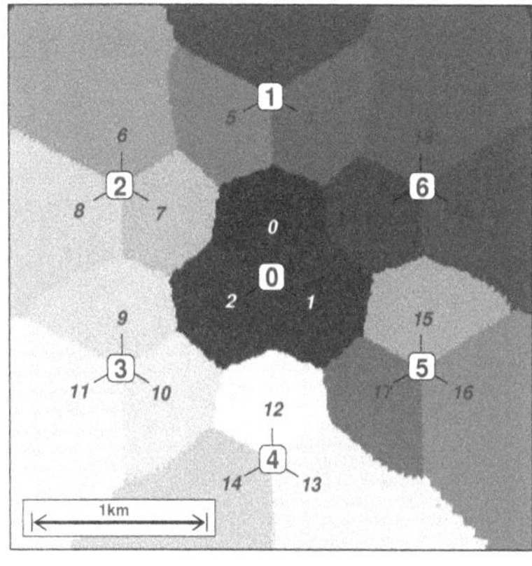


a) Best server with network structure

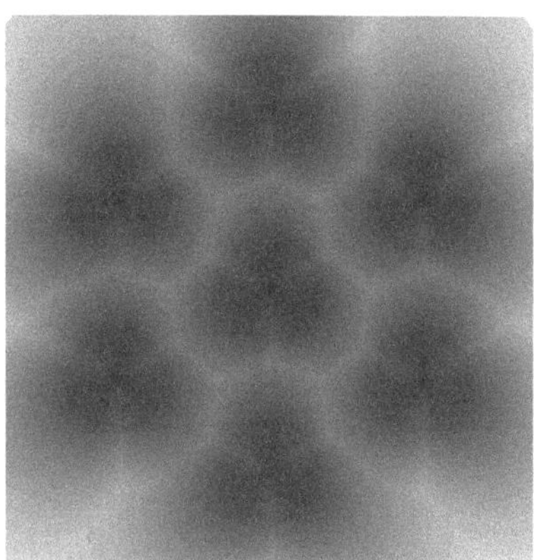

b) Minimum pathloss

**Figure 6.1:** Regular hexagonal scenario, 7 sites

The left part, figure 6.1 a), shows for each pixel the colour representing the cell to which the position has the best possible connection, the so-called best server. The cell structure is reflected by this view. The right part, figure 6.1 b), shows the minimum pathloss for each pixel. Areas close to the antennas have smaller pathloss and are darker. The lightest areas indicate cell boundaries meaning that beyond that boundary the pathloss is smaller to another cell.

### 6.1.4 Realistic scenario

Similar to the regular hexagonal scenario, figure 6.2 shows a realistic scenario of Berlin which consists of 143 cells. The scenario is publicly available and has been setup within the IST project *MOMENTUM*, see [Mom03]. The left part, figure 6.2 a), shows the best server plot reflecting the cell structure. The right part, figure 6.2 b), shows the minimum pathloss for each pixel. Some regions are a little darker than the surroundings, indicating areas with water on which the pathloss is small compared to other surface types.

## 6.2 UMTS simulation concept

According to the concept described in chapter 5, in the STD simulation concept for the application to UMTS, a simulation is started with an initial snapshot. This system state will be changed dynamically within the event driven part of the simulation until the STD window terminates and another snapshot is generated, independent of the previous initial snapshot.

Based on this concept, see [TPL+03] and [PLG+04], the simulation toolkit MoRaNET (**Mo**bile **Ra**dio **N**etwork **E**valuation **T**oolkit) has been developed within the framework of the IST project MOMENTUM. It has two modes, the *static* and the *STD* mode. The STD mode, see [LPTG03],

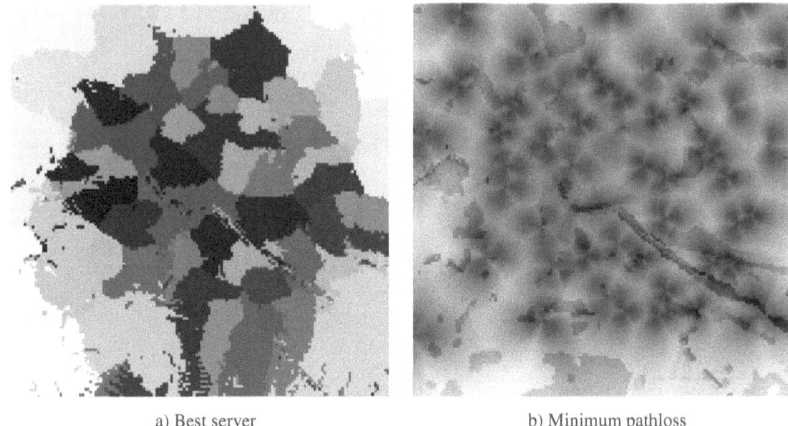

a) Best server b) Minimum pathloss

**Figure 6.2:** Realistic scenario Berlin

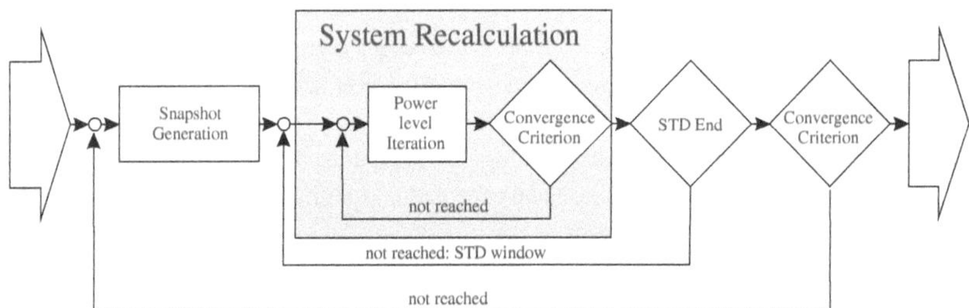


**Figure 6.3:** STD loop with system recalculation

is as described above and has been developed and implemented during this work. In the static mode, only random snapshots are generated, independent of each other. No event driven simulation is performed which would apply dynamic changes to a generated snapshot. A time axis is not present, the series of independent snapshots only uses a counter. The static mode is according to the left part of figure 5.1.

In figure 6.3, the major loops of a UMTS related STD simulation with MoRaNET are shown. The most inner loop takes place inside the iterative algorithm which performs the system recalculation, and it represents the *process input* action in the static part of figure 5.1. The loop ends when the iteration has converged according to a certain convergence criteria. The goal of this procedure is to reach a state in which all mobile stations of a cell send and receive on power levels which take not only the position relative to the base station but also the positions and power levels of surrounding mobile stations into account. Since the mobile stations influence each other, this calculation cannot be conducted within a single-step. For a more detailed description of this concept, see [TPL+03].

In the intermediate loop, the dynamic changes to the system, corresponding to the *generate new dynamic input* action in the short-term dynamic part of figure 5.1, are applied, and the loop continues until the end of the STD window is reached. An overall convergence criterion is checked in the outer loop. This can be an error calculation for a certain indicator function consisting of one or more system parameters, or it is simply the end of the last of a fixed number of STD windows. If the simulation end is not reached, another STD window starts after a new initial snapshot has been generated. This new snapshot corresponds to the *generate random input* action in the static part of figure 5.1. In figure 6.3, the loop outside the STD windows is included, while figure 5.1 is only related to a single STD window.

The static input data and the dynamic changes are described in the following. Most of the position oriented information is managed with the help of pixel maps. The size of the pixels is constant, but all pixel maps can have different pixel sizes. For more details on static scenario data as buildings, surface information, antenna positions and properties, etc., it is referred to the corresponding project documents [Mom03].

### 6.2.1 User profiles

A user profile is categorised by the service type. It specifies which service the user is going to use with a call. The different service types with their traffic source descriptions are listed in section 6.2.4. Popular user profiles are *SpeechTelephony* and *WWW*.

A user profile is connected to a list of so-called *usage types*. A usage type contains the information which *mobility type* it belongs to. The reason for having such a list for each user profile is that some combinations of service and mobility type are not reasonable, e. g., in the default configuration, the service type *VideoTelephony* is not possible together with the mobility type *highway*. The concept, however, is flexible, and if the person using the service sits in the back of the car while a driver is driving, the combination can be configured to be possible. Furthermore, if the requirements and possibilities change, highway users can, e. g., even be enabled on water environment to simulate speed boats. On the other hand, the restrictions which are configured make it possible to detect errors in the map configuration before performing lengthy simulations.

For each user profile which is present in the considered model, two traffic maps are required. The first is the *average load grid* (ALG) which represents the profile's stationary user distribution by determining for each location the mean number of users present of the profile. The second is the *busy hour call attempts* grid (BHCA) which determines the mean number of new users which emerge at the location within one (busy) hour.

For static users, i. e., non-moving users, the ALG can be easily calculated from the BHCA grid by using the mean dwell time for each pixel. For every pixel, Little's law applies:

$$N = \lambda T \tag{6.1}$$

In the equation, $N$ is the mean number of objects in a system, $\lambda$ is the mean number of objects arriving per time unit to the system, and $T$ is the mean number of time units an object dwells in the system. Applying this to the pixels, the system is the pixel itself, $T$ is the dwell time from the mobility map, see section 6.2.3, $\lambda$ is the arrival rate from the BHCA grid, and $N$ is the resulting value for the ALG for that pixel.

### 6.2.2 Operational environment

At the time a user is generated, the initial position within the scenario is chosen. This position determines the *operational environment* and by this the mobility type of the user. The operational environment map manages for each location a list of possible mobility types for each user profile which is possible at the location.

The operational environments which are used are the following: *water, railway, highway, mainroad, street, rural, suburban, open, urban, cbd (central business district).*

To give some examples, in the operational environment *street*, for the service types *Video-Telephony* and *StreamingMultimedia* only the *pedestrian* and *static* mobility is possible, while for other service types which can occur in that operational environment also the mobility type *street* is possible. On water, only *static* mobility is possible for all possible service types which can occur there. In *rural* environment, the penetration ratio of *static* vs. *pedestrian* mobility is $1:9$, in *urban* environment it is $3:7$.

The system, however, is flexible in the way that every possible combination of a list of service types, each with a list of mobility types and their penetration ratio, can be considered as an operational environment.

### 6.2.3 Mobility

A new pixel oriented mobility model has been introduced for UMTS network simulations, see [PEF+02]. This model takes care of keeping a user within appropriate areas, and it allows for modelling different traffic situations at different locations forcing users to change their behaviour based on their location. This enables, e. g., that a high-speed user can change his average speed when changing between highway, street and traffic jam environments.

With a mobility map, in which every part of the area has a set of parameters, these goals can be achieved. A part must represent a well-defined area. Easy to manage and to handle are maps in which every part is a regular area, optimally a square. Such a square is now called *pixel*, and for a single mobility map, all pixels are of the same size.

The parameters for a pixel consist of the mean dwell time for the pixel and the turning probabilities. Each pixel has 16 values for turning probabilities, for each of the four possible directions from which a user can enter the pixel, and for each of these incoming directions four values for

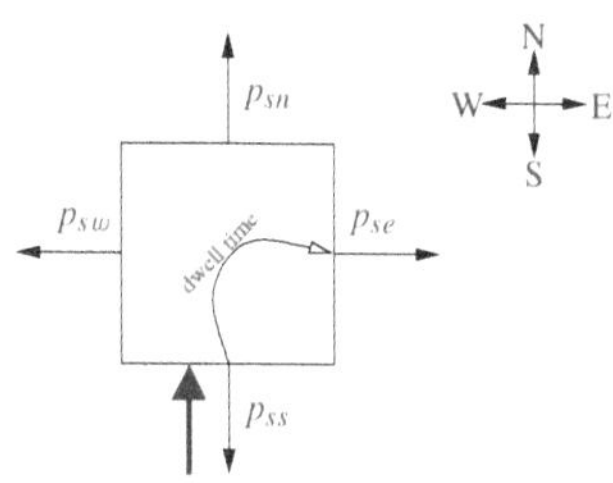


a) Single pixel from mobility map

| Probabilities in [%] | | To | | | |
|---|---|---|---|---|---|
| | | N | E | S | W |
| From | N | 0 | 0 | 0 | 0 |
| | E | 0 | 10 | 50 | 40 |
| | S | 0 | 45 | 10 | 45 |
| | W | 0 | 40 | 50 | 10 |
| New user | | 0 | 35 | 30 | 35 |

b) Example: turning probabilities for a north-bounded pedestrian mobility pixel

**Figure 6.4:** Single pixel from mobility map with example turning probabilities

the outgoing direction. Another four values are needed for users which emerge within the pixel and do not have an incoming direction. This results in 21 values for each pixel.

Termination probabilities for a pixel are not considered. An outgoing direction for a user which is located within the area represented by this pixel is only chosen if the user's dwell time for this pixel has elapsed while the session is still active, so that the user will leave the pixel. The point of time at which a user will not leave a pixel but terminate the session within the pixel, is always determined by the session duration and not by the pixel configuration.

With the turning probabilities, users can be restricted to appropriate areas, as mentioned above. At the scenario limits, e. g., all directions which would lead out of the scenario, are set to zero, and users will bounce back from the border or walk along the border, depending on the configuration. Also at the borders of operational environments, this technique can and should be applied. This is used to prevent, e. g., highway users from leaving the streets.

Figure 6.4 shows an example pixel configuration for a pedestrian mobility type. On the left side, figure 6.4 a), a pixel of the mobility map including a part of its parameters is shown. The situation displayed is for a user entering the pixel from the south, indicated by the thicker arrow at the bottom of the pixel. For such users, the probabilities to leave the pixel by one of the four directions are written next to the arrows at the corresponding pixel borders. The example shows a user coming from south and the leaving the pixel to the east after the dwell time has elapsed.

The table, figure 6.4 b), shows the values for the complete parameter set for the turning probabilities of the pixel. In this example, the pixel is north-bounded, meaning that it is not possible to leave the pixel to the north, no matter where the user has come from, and also impossible to enter the pixel from the north, if the maps have been configured correctly.

Since the mobility is quantised, the parameters only depend on the incoming direction and not on the exact position on the border. The arrow only means a user coming from the pixel adjacent to the south of the current pixel.

Only the mobility itself is quantised, however, not the user positions. Consider the following example: The pixel has the index (100,200), starting the map with (0,0) in the southwestern corner. The pixel size is 10 m, and thus, the pixel expands from (100 m,200 m) to (110 m,210 m). Now, a new user has been started in the pixel south to the displayed pixel, at position (101 m,197 m). After entering the displayed pixel, the user will be placed at the position (101 m,207 m). It is important not to put all users of a pixel at the same position in the center of the pixel. This would lead to unrealistic situations, especially regarding the radio conditions.

Since a pixel map is present for each mobility type in use, the speed of the users of this mobility type can be modelled heterogeneously within the areas possible for this mobility type. By this, a part of a street with high traffic jam probability can be configured with significantly lower average speed (higher value for the mean dwell time parameter) for the mobility type *street*. A street user passing through this area will automatically decrease the average speed at entering the area and again increase it after leaving. A pedestrian user, e. g., should have a lower average speed at crossroads, which also applies to the other non-static mobility types.

The mobility model reflects a short-memory model. The time a user has spent in the previous pixel is not taken into account at the determination of the actual dwell time in the current pixel. Also, the direction from which the user has entered the previous pixel is disregarded.

In contrast to static users, for moving users the calculation of the ALG from the BHCA grid (see section 6.2.1) is not sufficient to achieve an arbitrary but stationary user distribution. To illustrate this, an area is imagined where many users emerge according to the corresponding values of the BHCA grid, but from where the users move away with high probability, and to where users from surrounding areas move only with low probability. Even if the average dwell times for users passing this area would be comparable to the dwell times of other areas, the mean number of users in that area will be small.

For a steady state simulation as discussed in this work, it is mandatory that the number of users is stationary for every user type in every region. Therefore, the number of users per time unit entering the region must be equal to the number of users leaving. Since entering and leaving a region can be caused either by session events (start and end of a call) and mobility, calculation of the ALGs for non-static user types must take into account the corresponding BHCA and the mobility. See [PEF$^+$02] for a detailed description of the algorithm.

### 6.2.4 Session

The service type is assigned to a user profile and classifies it. The description of the service contains a list of radio bearers allowed for the service including information about statistical properties and requirements regarding the radio conditions when using the radio bearer. Further parameters deal with the up- and downgrading effects of the radio resource control (RRC).

Each service also has a description of the corresponding traffic source. It contains the information needed to generate events for the activity changes.

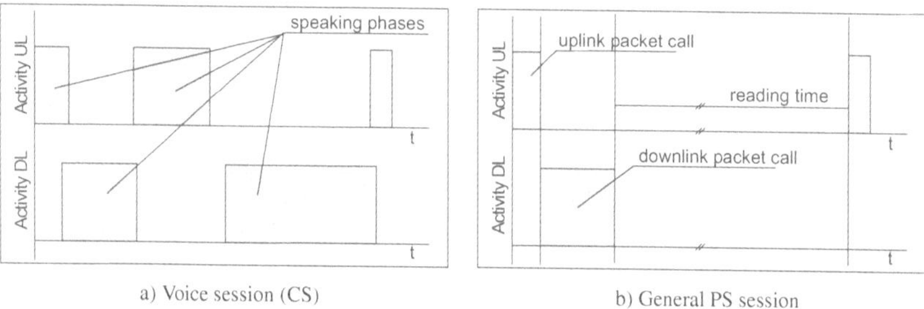


**Figure 6.5:** CS and PS sessions

In figure 6.5, a circuit switched (CS) session and a packet switched (PS) session are shown with respect to the phases of activity and inactivity.

The CS session in figure 6.5 a) is a voice session. Both uplink and downlink are used. In the uplink, the user is talking, in the downlink, the dialogue partner is talking. Pauses with inactive up- and downlink are also possible as are phases where the user and the partner are both talking. In a CS session, the involved link occupies the resources permanently. The difference between active and inactive phases is the lower interference contributed by an inactive link.

Active links contribute to the cell load, either by power (in DL) or by interference (in UL). This applies to CS sessions as well as to PS sessions. The uplink and downlink loads are defined as

$$\rho_{\text{UL}} = 1 - \frac{P_{\text{N}}}{I_{\text{total}}} \tag{6.2}$$

$$\rho_{\text{DL}} = \frac{P_{\text{total,DL}}}{P_{\text{max,DL}}} \tag{6.3}$$

According to [HT00], $P_{\text{N}}$ is the background and receiver noise, and $I_{\text{total}}$ is the total received interference at the receiver in uplink. In downlink, $P_{\text{total,DL}}$ is the total downlink transmission power of the base station, and $P_{\text{max,DL}}$ is the maximum transmission power.

Figure 6.5 b) shows a general PS session. The simulation is performed on the level of packet calls. In the configuration of a PS service, the distributions of the number of packets per packet call, and of the length of the packets in data units are specified. Further, the number of packet calls per session and the time between packet calls are specified. This results in a distribution of the amount of data per packet call. At the time of generation of a PS session, the number of packet calls and for each packet call the amount of data is determined.

The major difference between CS and PS sessions is that the length of a CS session and of its activity phases is time oriented, while in a PS session, it is data oriented in the sense that the determined amount of data has to be transmitted until the session terminates. The time needed for this transmission depends on the traffic situation. If there is a lot of traffic and the cells

| Bearer [kbit/s] | 384 | 128 | 64 | 32 |
|---|---|---|---|---|
| FileDownload | 0.64 s | 1.91 s | 3.82 s | — |
| WWW | 18.43 s | — | 30.57 s | 45.14 s |

**Table 6.1:** Traffic dependent PS session duration

serving the PS user is highly loaded, the user is only allowed to transmit at a lower data-rate resulting in a delayed session end. For CS services which can be transmitted with different data-rates, a lower data-rate only leads to lower quality of the transmitted data. The speech quality in voice services and the image and sound quality of streaming services will be decreased, but the session end is not delayed. Data oriented CS services are not included in this concept and will not be considered.

A problem resulting from this fact is the difficulty to calculate the ALG from the BHCA grid in cases where the session durations of PS services are traffic dependent. This problem is eased by the fact that load contribution of a PS user transmitting at a lower data-rate is lower than at a higher data-rate, and the load is the most fundamental basis for the decisions of the system about up- and downgrading, soft-handover, etc. This effect, however, is non-linear, meaning that two PS users transmitting at half the usual data-rate together do not have the same load contribution as a single PS user transmitting at the usual data-rate.

As an example, the *WWW* service is considered. Is has a mean number of packet calls per session of 5, geometrically distributed. Also geometrically distributed is the number of packets per packet call with the average of 26. A single packet has a Pareto distributed length with a mean of 896 byte. The negative exponentially distributed reading time between packet calls is 4 s. This configuration of the service has been taken from the UMTS standard [ETS98] and the project documents [Mom03], as has been done also for all other services in simulations considered in this thesis. Assuming now the highest possible data-rate using a radio bearer with 384 kbit/s during the entire session leads to an average session duration of about 18.43 s of which 16 s are pure reading time. Table 6.1 shows for the *WWW* and the *FileDownload* service the mean session durations for the three radio bearers which are possible for each service.

The services can all be freely designed with respect to the distributions for packets, packet calls, and the reading times for the PS services, and to distributions of on- and off-phases for the CS services. The configurations considered in this thesis are – as mentioned above – taken from the UMTS standard and the project documents, see [ETS98] and [Mom03]. The used CS services are *SpeechTelephony* (the voice service), *VideoTelephony* and *StreamingMultimedia*, the used PS services are *EMail*, *LBS* (location based services), *MMS*, *FileDownload* and *WWW*.

Further possible important differentiations between the services are link restrictions, e. g., only up-, only downlink, or selective, i. e., either up- or downlink but not both for the lifetime of a session. Possible data-rates and configuration regarding Quality of Service aspects are specified. The latter includes parameters for the radio resource management (RRM) which decides about actions like up- and downgrading, blocking, and dropping of calls. These actions can be classified and weighted to make optimisation possible for different goals like highest possible utilisation of the radio network, lowest possible ratio of dissatisfied users due to blocking or dropping of calls, etc.

### 6.2.5 User generation

Each user type has its own module responsible for the generation of users of this type. Two maps are needed for the user generation, as mentioned in section 6.2.1. The first map, the ALG, is used at the beginning of an STD window for the initial snapshot. The number of users of this type is drawn which will be in the initial snapshot. Next, these users are distributed over the scenario area according to the map.

During the STD window, newly arriving users are generated one by one. Not the number of users is drawn, as in the beginning, but the inter-arrival time for the next user of this type. This single user is placed onto the scenario according to the BHCA map which is similar but not exactly equal to the ALG.

#### 6.2.5.1 Classification process

For both, the initial generation of users and the single user generation, the position of each generated user is determined. According to the pixel this position belongs to, the operational environment (OE) is determined. This information is needed to determine the mobility type of the user.

Assigned to the pixel of the operational environment map is the index of the OE. Each OE has a configuration with a list of user profiles which are allowed within this OE, or it has a single entry without specification of the user profile meaning that the entry is common to all user profiles. It is an error if a user of a certain profile has been put to a pixel where the OE has entries specifying the possible user profiles and the user profile corresponding to the user is not in this list.

An entry in the configuration of an OE specifies the possible mobility types, either general or for a specific user profile. The possible mobility types have a probability, all of which have to sum up to unity within a single configuration entry of an OE.

At this point, finally, the generated user has, additional to its user profile, the position where to start, and the mobility type. The mobility type cannot be changed during the session (lifetime) of a user, but the speed and operational environment can be changed.

To recall this procedure, the following enumeration shows the order of decisions made when generating a user:

1. All user profiles used in the scenario are selected sequentially.
2. For the selected user profile, a group of users or a single user is generated.

3. For each generated user, the position in the scenario area is decided.
4. The operational environment is determined at the pixel to which the position belongs.
5. The general or user profile specific list of possible mobility types is selected.
6. From that list, a mobility type is randomly selected, taking the probabilities for the mobility types into account.

#### 6.2.5.2 Event generation

After the user has been placed and configured, the events reflecting the dynamic aspects of the user have to be generated according to the user profile and the mobility type.

First, the duration of the session is determined. According to this, all mobility events, i. e., location changes from one pixel to another, are pre-generated. The duration of certain session types can be extended due to downgrading, i. e., a given amount of data in a PS session needs more time for transmission if the data-rate is lower than requested. For these session types, mobility events are generated for a longer duration. This means the worst case assuming the lowest data-rate. The pre-generation of events allows for simpler techniques to re-schedule pending events in case of up- or downgraded sessions.

Changes to the data-rate due to up- or downgrading are also a kind of activity change, but the corresponding events are generated during the simulation and not at the time of the user and session generation.

Up- resp. downgrading of the data oriented PS services require a recalculation of the pending activity changes since the time at which the current packet call will be finished has to be put to an earlier time resp. has to be postponed. Up- and downgrading during an inactive phase of a PS session does not occur since the session is in a state occupying no resources.

## 6.3 Evaluation of dynamic QoS

A flexible design with a central evaluation controller makes it possible to implement code for evaluation probes at any point in the simulation source where data of interest can be gathered. The data can be provided to evaluation modules producing the appropriate statistics.

Many kinds of information can already be used and processed by a static simulation, but only static data can be collected that way. These are statistics about system related parameters like soft-handover probability for each pixel, or user related statistics like blocking probabilities for different services and areas of the scenario.

Soft-handover (SHO) is the case in which a call is connected to more than one cell. If these cells belong to the same base station, it is called softer-handover (SSHO). In this concept, softer-handover is a subset of soft-handover, meaning SHO includes also the SSHO cases. Soft-handover is described in more detail in section 6.4.

Quality of Service (QoS) as one of the most important performance aspects needs to be predictable to make it possible for providers to satisfy their customers and fulfil contracts with the least possible costs. Dynamic QoS parameters cannot be evaluated by static simulations, only static QoS like the blocking mentioned above.

For dynamic QoS, the dynamic aspects of the mobility and the session need to be considered, as it is done in the STD simulation concept. As one of the most popular examples, call dropping can only occur if the call has been accepted previously. Some changes in the environment of the dropped call cause the call to be dropped. Usually, the user has moved to an overloaded cell or an uncovered area, or the load situation of the connected cell has changed.

What distinguishes a dynamic QoS parameter from a static one is the fact that a history of the investigated parameter is required to evaluate it, and possibly also information about its future. In the case of dropping, only the single fact is needed that the call has already existed before. For parameters like the delay in PS sessions, all delays a call has experienced up to the current simulation time need to be accumulated, and for a cut session in case of an ended STD window, the planned end also needs to be taken into account.

Many statistics of different categories are possible, and the following lists show the most important ones which have been implemented and used.

### 6.3.1 Session based

Session based QoS statistics are those which evaluate a counter or a duration value on a per session basis.

**Soft-handover ratio:** The session based SHO ratio describes the sum of the durations of all phases in which a user has been in SHO, and this is put in relation to the total session duration.

**Soft-handover duration:** This is the time between entering a SHO area and leaving it again. It can happen several times within a session, and the times are considered separately without accumulation. If the mean value of these phases is taken and a distribution of these mean values is evaluated over all sessions, this is a session based statistics. It can also be a system based statistics, see section 6.3.2.

**Softer-handover duration:** Similar to the soft-handover duration, this statistics considers the softer-handover. This can also be evaluated system based.

**Dropping:** The probability for a user that his session is disconnected during the session lifetime.

**Downgrading:** Several statistics are related to down- and upgrading. The relative usage of the radio bearers possible for the service can be evaluated on a per session basis. The distribution of these ratios over all sessions can, in turn, be evaluated. Another session based statistics is counting the users who have been downgraded during the session and calculating the downgrading ratio. A user who has not been given the best possible radio bearer at the start of the session is also classified as being downgraded, and the difference between the assigned data-rate and the best possible data-rate can be evaluated. These users can also be evaluated with the static simulation, while this is not possible for the users who have been downgraded during their sessions.

**Delays:** Delays can only occur in the data oriented PS sessions. The distribution of the delays is a possible statistics, and further, the ratio of delayed sessions. The latter is closely related to the ratio of downgraded sessions, but CS sessions can also be downgraded. According to the statements further down in section 6.3.3, reasonable statistics on the delay will be service type restricted. Figure 6.6 shows two examples.

**Cell changes and visits:** During a session, a user can change the reference cell, i. e., the best cell in the active set. The statistics about cell changed counts all changes during a session. A cell change can be, e. g., in SHO changing from an active set *A,B* with *A* being best cell to an active set *B,A* with *B* being best cell. Also leaving SHO by changing the active set from *A,B* to *B* is a cell change. In contrast to this, leaving SHO by changing from *A,B* to *A* is not counted here. The number of cell changes can be 0, especially for non-moving users. In the statistics on cell visits, each cell which has been best cell at least once during the session, will be counted only once, and thus, the minimum number is 1.

Session based evaluations have to cope with the possibility of having cut a session of which information has been gathered. A cut induced by the system is fine since it does not bias the evaluation, but a cut induced by the simulation at the end of an STD window either makes the affected session unusable for evaluation, or some kind of realistic prediction is needed to make the session usable. It is up to the simulation operator whether sessions are disregarded which already existed as active session in the initial snapshot, and whether sessions are disregarded which have not finished before the end of the STD window, i. e., the cut sessions.

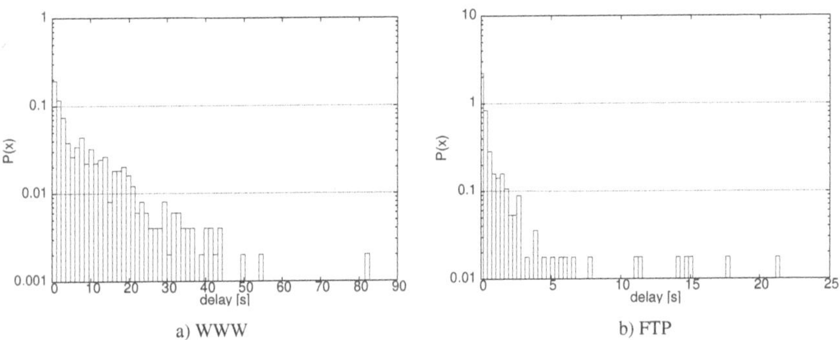


a) WWW b) FTP

**Figure 6.6:** Delay histograms for PS sessions

Figure 6.6 shows the session delays for the realistic scenario. The session delay is not related to transmission delays, but it is defined as the difference between the actual session duration and the time the session would have lasted if the MS could have transmitted always at the highest data-rate. Such a session which has started with the highest data-rate and has never been downgraded, thus, has a zero delay per definition. Delays are absolute and given in units of seconds, and so sessions with higher data volume are more likely to suffer higher delays. In figure 6.6 a), the delays for the WWW users are shown, and in figure 6.6 b), the delays of the FileDownload users are shown. Only new and completed sessions have been evaluated disregarding the sessions that have been active at the initial snapshot or that have not been finished before the end of the STD window, see end of section 6.3.1.

### 6.3.2 Location or system based

Location oriented QoS statistics are those which evaluate counters or ratios of parameters at pixel level without paying attention to the rest of the session which the obtained value has been taken of.

**Soft-handover ratio:** The location oriented SHO ratio is at the first sight the same as the static location oriented SHO probability. The dynamic version of the SHO algorithm, however, takes into account the hysteresis parameters, and this can have some impact on the ratio especially at the borders of SHO regions.

**Softer-handover ratio:** Similar to the soft-handover ratio, but this statistics is about the softer-handover.

**Active set size** This evaluates the location oriented (mean) active set size for users at that pixel. The statement from above about the static nature of these statistics and the difference in applying the dynamic SHO algorithm applies to all static SHO statistics.

**Branch events:** These events indicate a change of the composition of the active set of a user. Either a cell is added, a cell is deleted, or a cell is replaced by another cell. These different events can also be evaluated separately. Evaluation of these events is impossible with static simulations since either the user has moved to another area making the consideration of mobility necessary, or the radio conditions at the borders of SHO regions have changed while the user remains motionless.

**Blocking:** This is the probability for a new user to be blocked at the pixel the user was planned to start the session. This is very close to a simple static statistics, but while in a static simulation all users in the system are treated equally, in the STD simulation a new user is blocked in favour of already existing users in high load situations. In static simulations, among those users whose blocking would resolve the high load situation, one user is chosen randomly. In STD simulation, it will be the new user who is blocked. In static simulation, among the generated users some can be classified as new users while the others are treated as existing users. This enables the preferred blocking of new users and even the distinction between blocking and dropping, but only on a statistical basis including a-priori information to classify the users as being new or already active ones.

System oriented statistics can be, e. g., on cell level or on system level. As stated in section 6.3.1, the SHO duration can be evaluated for a certain cell without considering individual sessions or the exact pixel position.

### 6.3.3 Common methods

All statistics described in sections 6.3.1 and 6.3.2 can be restricted to certain parameters. The most important restrictions are the following:

**Service type:** Only obtained values for which the user profile of the corresponding user matches a given service type are taken into account for evaluation.

**Service class:** As for the service type, the restriction can be on the service class which is less strict. Service classes are *conversational*, *streaming*, *interactive* and *background*.

**Cell:** Only users whose reference cell matches the given (list of) cell(s) are considered. For session based statistics, the cell restriction is intuitively important, e. g., finding the dropping rate in a certain cell can influence the decisions on resource dimensioning for the corresponding base station. On the other hand, the kind of location orientation introduced by cell oriented statistics is already included in the location oriented statistics and even with finer resolution, but a pixel can belong to more than one cell in SHO situations. It can in any case be interesting, however, to have the statistics on cell basis.

### 6.3.4 Detailed radio resource management

In the STD concept it is also possible to examine some more detailed protocol behaviour. One example is the radio resource management (RRM) with focus on the call admission control (CAC) and congestion control (CC). This has been implemented as an optional feature which does not need to be activated. Essential actions controlled by the RRM are blocking, dropping, up- and downgrading.

An important aspect to adhere to is the realistic modelling of the availability of information to the different model components involved. In the basic mode without the RRM extension, the decisions about the RRM actions are based on the *measured* cell load. This parameter, however, is not available at the entity which makes the decision, but has to be estimated instead. Solutions for this can be integrated into the simulation toolkit according to the planned solutions of the network hardware manufacturer or the network operator.

The assumptions which the estimation mechanism is based on can be evaluated by the simulation to which the real load values are available. The estimation mechanism itself can be tested. E. g., static and adaptive approaches can be simulated. In the static approach, all parameters for the estimation mechanism are defined prior to the simulation run, and they keep their values unchanged. Adaptive approaches can make use of the fact that there is some instance in the network knowing the real values. Regularly, these values can be communicated to check the quality of the parameters of the estimation mechanism.

Even more advanced approaches can try to detect a relation between parameters indicating the current network situation and the best parameters for the estimation mechanism for such a network situation. Information gathered by such investigations can be used to optimise less advanced or even the static estimation mechanism.

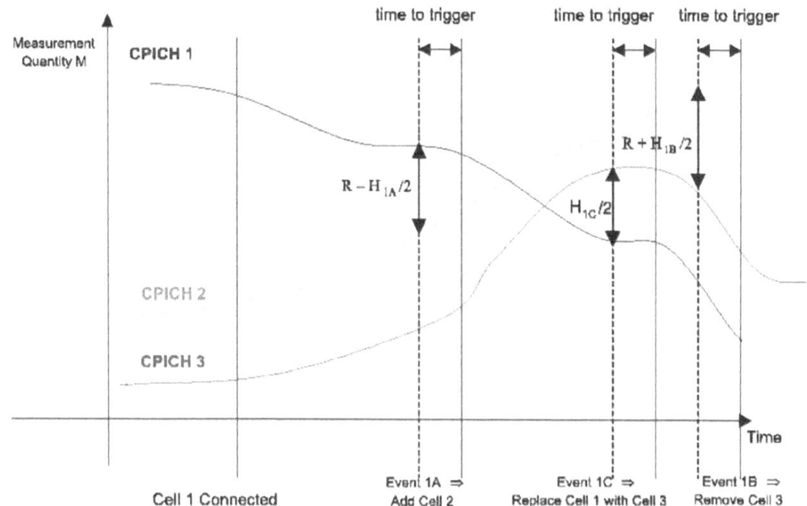


**Figure 6.7:** Handover events: Addition (1A), Replacement (1C), Deletion (1B)

## 6.4 Soft-handover

Soft-handover (SHO) is an algorithm of the (W)-CDMA standard where an MS can be simultaneously connected to two or more cells during a call. The purpose is to have seamless handover while moving, without having a noticeable service disruption. In contrast to 2G systems as GSM which follow the *break before make* strategy by disconnecting the current cell before connecting to the new cell, in 3G systems as UMTS, the *make before break* strategy is followed. The connection to the new cell is established before the connection to the old cell is released.

Another advantage of SHO is the link gain. On the one hand, there is the diversity gain that comes from the reception of the same signal from more than one transmitter (in downlink). Similarly, in the uplink more than one base station receives the signal. In both cases, the required transmission power for a certain signal quality is lower in the SHO if the pathloss difference between the connected base stations is not too high. For details see [HT00].

### 6.4.1 Basic Principle

Each link between the MS and the base stations involved in a SHO are referred to as a branch. Continuous measurements of the strength of pilot carrier transmissions (Common Pilot Channel - CPICH) between the MS and the neighbouring cells are carried out.

Figure 6.7 shows the variation over time of three such measurements (CPICH1, CPICH2, CPICH3) corresponding to an MS. These measurements are referenced in the following as $M_{new,\mathrm{dBm}}$, $M_{best,\mathrm{dBm}}$ and $M_{old,\mathrm{dBm}}$, depending on the branch event.

The main parameters of the SHO algorithm which are involved in deciding the handover events, are the Reporting Range ($R$), the Addition/Deletion Hysteresis ($H_{1a}$, $H_{1b}$) and Replacement Hysteresis ($H_{1c}$), specified in *dB*. The three branch events *radio link addition* (RLA), *combined radio link addition and removal* (CRLAR) and *radio link removal* (RLR), are introduced in the following. They are also called *branch addition*, *branch replacement*, and *branch deletion*.

#### 6.4.1.1 Radio link addition

In the example situation shown in figure 6.7 originally the MS is connected to cell 1. Once the condition for Radio Link Addition (RLA) is satisfied according to equation (6.4), a link is established between the MS and cell 2, making the cell a member of the active set of the MS (event 1A in figure 6.7):

$$M_{new,\mathrm{dBm}} \geq M_{best,\mathrm{dBm}} - (R - H_{1a}/2) \tag{6.4}$$

In figure 6.7, $M_{new,\mathrm{dBm}}$ is CPICH2 and $M_{best,\mathrm{dBm}}$ is CPICH1 for this event.

#### 6.4.1.2 Combined radio link addition and removal

A branch replacement (event 1C in figure 6.7) takes place when the active set of the MS has reached its maximum size and the strength of a CPICH decreases and becomes the worst while another CPICH gradually increases leading to satisfaction of the condition for Combined Radio Link Addition and Removal (CRLAR):

$$M_{new,\mathrm{dBm}} \geq M_{old,\mathrm{dBm}} + H_{1c}/2 \tag{6.5}$$

In figure 6.7, $M_{new,\mathrm{dBm}}$ is CPICH3 and $M_{old,\mathrm{dBm}}$ is CPICH1 for this event.

#### 6.4.1.3 Radio link removal

When the signal of a cell decreases significantly leading to satisfaction of the condition for Radio Link Removal (RLR), then cell 3 is deleted from the active set (event 1B in figure 6.7):

$$M_{old,\mathrm{dBm}} \leq M_{best,\mathrm{dBm}} - (R + H_{1b}/2) \tag{6.6}$$

In figure 6.7, $M_{old,\mathrm{dBm}}$ is CPICH3 and $M_{best,\mathrm{dBm}}$ is CPICH2 for this event.

The conditions for the branch events contain parameters which are a subject for optimisation regarding the system performance in a process of network dimensioning.

The reporting range $R$ specifies a gap between the received signal quality (regarding the common pilot channel) of two cells. If the signal difference is lower than the gap, a new cell is considered to be included into the active set. Additional hysteresis margins for adding, deleting and replacing a cell avoid fast in and out actions reducing the signalling traffic between the base station and the radio network controller (RNC). The parameters influence the total soft-handover ratio (the percentage of users in soft-handover), the soft-handover durations per user, the cell sizes and the hardware utilisations in the access network.

### 6.4.2 Static and dynamic soft-handover simulations

In static simulations, where only a snapshot of the system is available, only the reporting range can be used to determine the members of the active set. The branch events for which the conditions in the equations (6.4), (6.5) and (6.6) have to be fulfilled, do not actually happen. E. g., at the time when the event 1A happens in figure 6.7, to a static simulation only the current measurement values for the three pilot channels are available. Therefore, all cells which $M_{\mathrm{dBm}}$ have a distance less than $R$ to the $M_{best,\mathrm{dBm}}$ become members of the active set, up to the maximum number allowed for an active set.

In dynamic and STD simulations, the incorporation of history information makes it possible to consider the difference between the current situation and the previous situation. Especially the mobility of the users can be taken into account. Mobility is the main reason for changes needed in the active set of the user.

In figure 6.7 at event 1A the cell with the CPICH2 can be identified as a cell which has been absent from the active set before and is now approaching the CPICH1. As a cell which is considered for addition to the active set, equation (6.4) can be applied using the addition margin $H_{1\mathrm{a}}$. The CPICH of a cell considered for addition has to be by $H_{1\mathrm{a}}$ closer to the cell with the best pilot level than the reporting range $R$. This makes it more difficult for a cell to be added. Similar at the event 1B in the figure, the cell with the CPICH3 is known to having been member of the active set, and it is considered for deletion from the active set. Knowing this, the SHO algorithm uses equation (6.6) including the deletion margin $H_{1\mathrm{b}}$. All three branch events cannot be realistically simulated with static simulations that need to use a SHO algorithm which is different from the (short-term) dynamic SHO algorithm introduced in section 6.4.3.

A further possibility of the dynamic and STD simulation is to implement the *time to trigger* feature. By this, the execution of a branch event can be made dependent on the time elapsed since the conditions for the branch event became true. The number of branch events can be reduced by this mechanism, which reduces signalling traffic.

### 6.4.3 Algorithm

In the soft-handover algorithm for a single MS (figure 6.8) the so-called input active set is determined. It is used to tell the simulation not to generate the active set with the static method. The algorithm is conducted prior to every system recalculation.

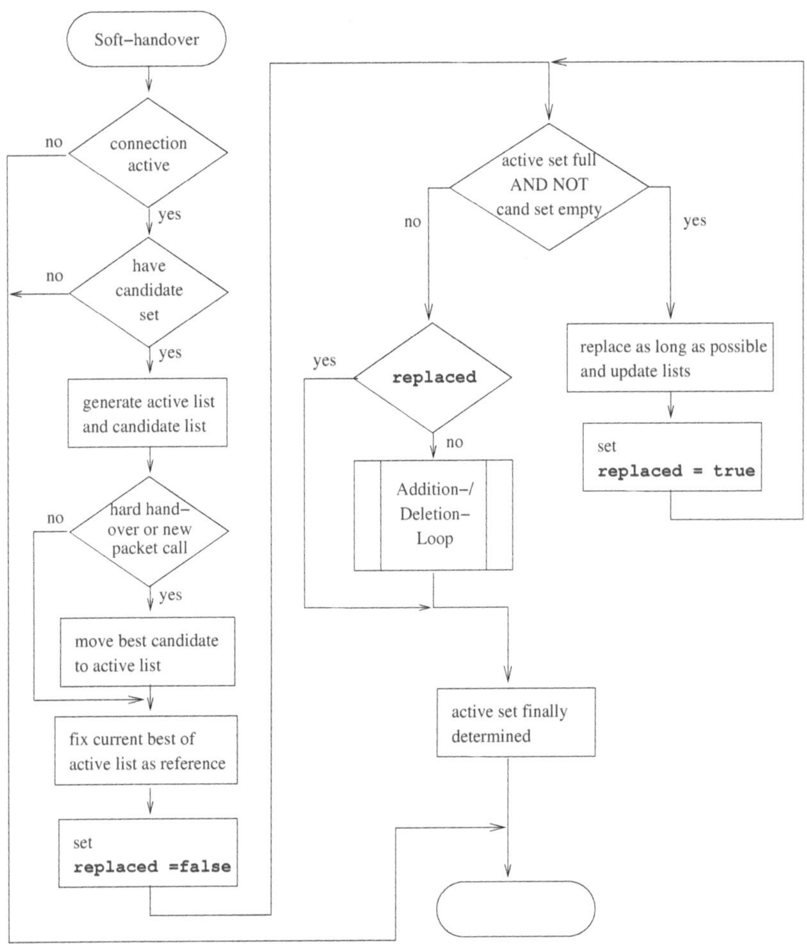


**Figure 6.8:** Soft-handover algorithm

It is first checked, whether the connection is active. This only concerns inactive PS connections (between packet calls), since in this concept, users with CS sessions only exist while their session is active. The next is the check for a candidate set. If the mobile station does not currently have a candidate, it is in outage and not considered further by the algorithm.

The next step is to consider the intersection of the current active set with the candidate set. This will form an *active list* as an initial set for the decisions on adding, deleting and replacing. The *candidate list* is the list of candidate cells without the cells included in the *active list*.

If the *active list* is empty, either a hard handover has happened or a new packet call (PS service) has started (during the inactive reading time there was no active set). In that case, the best of the candidate set is put to the active list to form a non-empty initial set.

For upcoming decisions which are based on cell specific configurations, the current best server is fixed to use it as reference. This is necessary, since later after adding some cell, the *real* best server could be changed.

Next it is checked, whether a replacement is possible. This is the case only if the active set, i. e., the *active list* in this context, is full, meaning it has reached its maximum size. In this case, replacements are conducted as long as the corresponding conditions hold, see equation (6.5).

Only if no replacement has taken place, the addition-deletion loop is entered. In this loop, it is tried first to add as many cells as possible according to the conditions, next the same will be tried with deletions. Since it is possible that after one or more deletions again one or more additions are possible if the size of the active set is less than the maximum, the loop is executed once again, if a deletion has happened before. This will be repeated until convergence, i. e., no deletion has happened within the last loop. A possible objection to this behaviour could be that a deletion followed by an addition could also be done by a single replacement. From the system view, however, this is not the same, and the parameters $H_{1a}$, $H_{1b}$ and $H_{1c}$ can be configured in such a way, that combinations of additions and deletions are preferred against replacements or vice versa.

After going through the loop, the active set is considered to be finalised from the view of the soft-handover algorithm.

### 6.4.4 Simulations

Examples of simulation results are shown in table 6.2 and in figure 6.9. The default settings have been 1 dB for the margin parameters $H_{1a}$, $H_{1b}$, and $H_{1c}$, and the default reporting range has been $R = 4\,\text{dB}$. The simulated models have been a regular scenario with 19 sites (base stations) and 3 sectors (cells) for each site, resulting in 57 cells, and the realistic scenario shown in figure 6.2. In the regular scenario, two mobility types, namely *mainroad* and *pedestrian*, have been simulated separately. They differ in average speed (54 km/h vs. 3 km/h) and in the movement behaviour, see section 6.2.3. E. g., the tendency to keep the current direction is higher for mainroad users than for pedestrian users.

| mobility type | ratios / % | | mean time / s | | mean | mean # | | mean # BEs / s | | |
|---|---|---|---|---|---|---|---|---|---|---|
| | SHO | SSHO | SHO | SSHO | AS | CV | CC | BA | BD | BR |
| pedestrian | 24.8 | 7.6 | 50.7 | 44.1 | 1.31 | 1.01 | 0.08 | 0.78 | 0.94 | 0.03 |
| mainroad | 24.1 | 7.8 | 10.6 | 7.0 | 1.30 | 1.39 | 0.82 | 18.5 | 19.2 | 0.76 |

*AS:* active set size, *CV/CC:* cell visits/changes, *BEs:* branch events, *BA/BD/BR:* branch additions/deletions/replacements

**Table 6.2:** Soft-handover characteristics of voice users, regular scenario

With the default parameters, the regular scenario provides the following results (table 6.2): About 24 % to 25 % of the users are in soft-handover (SHO), whereas 7.6 % to 7.8 % are in softer-handover (SSHO). The *mean time [s]* column represents the soft(er)-handover duration as described in section 6.3.1. The table shows the mean values for both mobility types. These mean values are higher for the slower pedestrian users than for the faster mainroad users.

The mean active set size is almost the same for both mobility types, while the number of cell changes and visits (see section 6.3.1) is much higher for the faster users, as it is expected. The higher number of branch events per time unit (see section 6.3.2) for the high speed users also corresponds to the higher number of cell visits and changes.

In figure 6.9, the SHO and SSHO ratios are plotted over the reporting range $R$, and the mean active set size is plotted over the deletion margin $H_{1b}$. The above mentioned default parameters have been fixed while in figure 6.9 a) the reporting range and in figure 6.9 b) the deletion margin has been varied.

The reporting range parameter has the highest influence on the system performance. An increased reporting range not only increases the SHO ratio by including relatively weaker connections, it also increases the soft-handover areas since mobile stations further away from the base station can be included. As a consequence, the mean active set size increases.

Since the SHO ratio in this example changes very similarly to the mean active set size, in figure 6.9 a) only the SHO ratio is shown, and in figure 6.9 b) only the mean active set size.

In figure 6.9 a), the SHO ratios and the SSHO ratios increase almost linearly for all scenarios. Approximately one third of the mobiles in soft-handover are in softer-handover for the regular scenario. For the realistic scenario, the number of users in soft-handover is quite large, causing the SSHO figure to be much lower at about one fifth. This fraction increases with increasing reporting range for the regular scenario whereas no such increase is observed for the realistic scenario. This effect is explained by the fact that in realistic scenarios both the soft- and softer-handover regions are irregular, see figure 6.2 b).

The impact of the three margin parameters is shown exemplarily for the deletion margin in figure 6.9 b) in the range from 0 dB to 2 dB. The influence on the mean active set size is shown.

The addition and deletion margins affect the mean active set size and the SHO ratio antithetical. While an increased addition margin delays a user entering an SHO situation and by this shortens

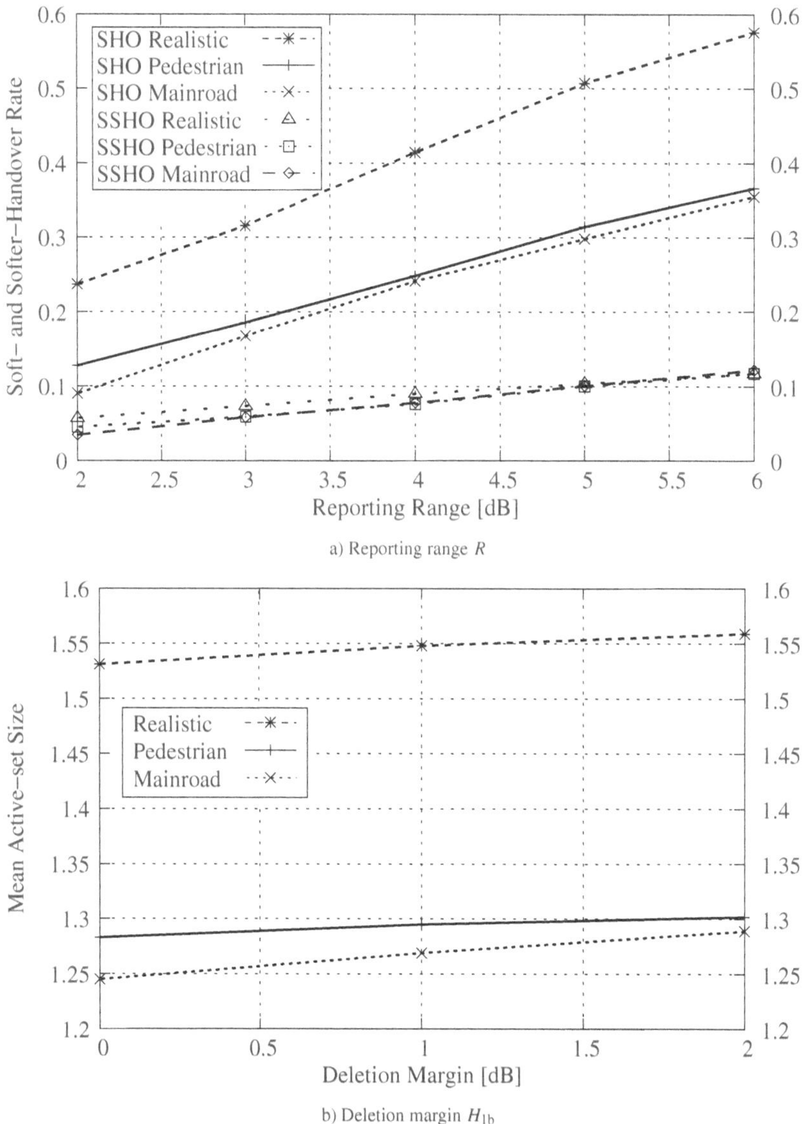


**Figure 6.9:** Variation of reporting range and deletion margin

the SHO region, an increased deletion margin delays a user being in SHO releasing a cell from the active set, and by this it stretches the SHO region. Resulting from this, increased addition hysteresis prevents some users from entering SHO at all, leading to a decreased SHO ratio. Increased deletion hysteresis increases the SHO ratio and consequently the mean active set size. These effects are more dominant for users moving at high speed (mainroad) than for the lower speed pedestrian users, see figure 6.9 b).

## 6.5 STD performance

For the investigation of the performance of STD simulations, quantitative statements about the development of the relative error need to be investigated. These error measurements are based on the LRE algorithm, see section 2.4, taking into account the local correlation of the evaluated sample. Reaching a certain error level is one event which can be defined as having reached convergence. To recognise this convergence, an appropriate parameter is needed. The real-time evaluation of this parameter can be used as a convergence indicator.

### 6.5.1 Choice of convergence indicator

The dynamic QoS parameters, see section 6.3, are usually counters or ratios. A major disadvantage emerging from this fact is that there is no sequence of values of which conclusions about the correlation can be drawn. This makes these parameters inappropriate for correlation oriented error observation.

Other parameters usually are of the kind *duration of X* or *time between* $X_1$ *and* $X_2$, or a distribution of such parameters. These parameters cannot be used as indicators with small STD windows, and as it has been shown in section 5.3, only for quite small STD windows a significant performance gain can be achieved.

Consequently, a parameter with instantaneous values is needed without evaluation focus. It means that since the dynamic QoS parameters of interest are inappropriate for this purpose, another parameter can be chosen which the simulation operator does not need to be interested in regarding a statistics about that parameter, but it is chosen only to indicate the simulation convergence.

The parameter of choice for the convergence status variable or the indicator, is the cell load. Apart from providing instantaneous values, the cell load has two properties qualifying it for this task. First, the effects about the dynamically changing amount of active users and active links are incorporated which is the most fundamental dynamic factor of the system. Secondly, a large sample size is provided since the observations occur very frequently, namely with each system recalculation.

The relation of the cell load to the system dynamics has already been introduced in section 6.2.4. The load of a cell depends on the number of active links corresponding to the number of users currently connected to the cell and currently having an on-phase in the corresponding link. In the uplink case of, e. g., voice sessions, it corresponds to actively speaking users.

Further parameters which the cell load depends on, are the positions of the users, mainly the positions relative to the antenna. This relative position consists of the distance, the angle, and the azimuth.

The cell load in uplink or downlink of a certain cell is only a single dimension of the system state which has a practically infinite number of dimensions. This indicator, however, is suitable to investigate the local correlation of the corresponding stochastic process representing a small part of the complete system behaviour. And if required, a combination of convergence indicators representing different areas of the scenario can be performed, i. e., with an AND relation.

### 6.5.2 Comparison with analytical investigation

In contrast to section 5.3 which showed an analytical examination of a simulation model, the purpose of this section is to show for simple simulation models which cause some parts of the simulation process to have Markovian properties, that the behaviour regarding the correlation is similar to the model used in the the analytical investigation in section 5.3.

#### 6.5.2.1 Similar model

To compare the results of the analytical investigation of section 5.3 with the results of UMTS simulations, the simulations in this section are focused on the speech model. A user changing from an on-phase to an off-phase in uplink (the user stops or pauses speaking) reduces the load in the uplink of the cell(s) to which the user is currently connected. The same applied to the downlink which is considered separately from the uplink.

Because of the quite large number of parameters forming the cell load, mainly every user's relative position to the antenna, the corresponding random variable is a continuous one. This is different from the random variable representing the occupancy in section 5.3 and which corresponds to the states of the Markov chain.

The number of active voice users with active uplink or downlink is, however, the major parameter contributing to the uplink or downlink load, and the duration of the on- and off-phases are negative exponentially distributed as well as the session duration. Thus, the variation of the uplink or downlink load over the simulation time reflects a stochastic process very close to a Markov process.

The states of the Markov chain would simply be the sum of active voice users connected to the cell and being currently in on-phase in uplink or downlink, depending on the link considered for this special process. In case of different intensity parameters for the on- and off phases, a state could be described by a pair of sums for the voice users in on- and in off-phase. This model would be possible to be calculated analytically, admittedly with more effort than the model used in section 5.3.

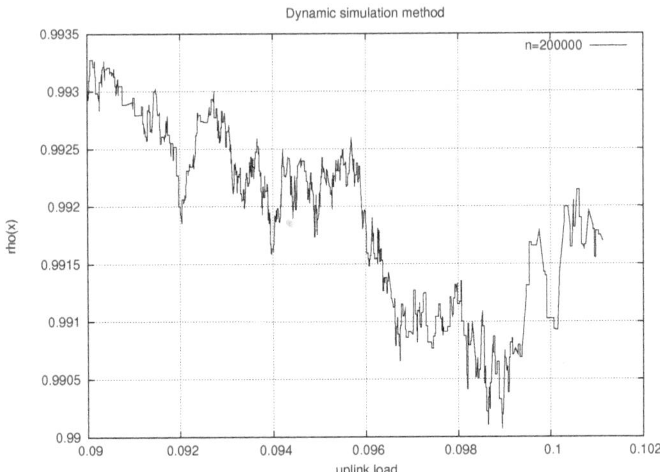


$n = 200000$: number of values for complete simulation

**Figure 6.10:** Local correlation vs. UMTS load, dynamic simulation

#### 6.5.2.2 Simulations

The uplink load is chosen here to represent the system state, and only one cell (index 0, one of the cells of the central site) is considered. The scenario used is the scenario described in section 6.1.3. 500 active users with voice service are distributed uniformly over the scenario. The variable plotted on the $x$-axis is always the uplink load, or the mean value of a group of uplink load values.

The ranges of the uplink load which are shown in the following figures are not chosen to show only extracts of the results. It always shows the complete range for which the simulation provided sufficiently accurate values with respect to the given error limit for all plots included in a single figure.

First, figure 6.10 shows the local correlation for a dynamic simulation part. This means, only a single STD window has been evaluated, or more precisely, the first 200,000 values of an STD window have been evaluated. The correlation is very high, the variations are very small considering the very small range on the $y$-axis. It corresponds to figure 5.9, in which the model even with the highest load (the utilisation of the M/M/1 model) remains far below the correlation in figure 6.10.

Next, figure 6.11 shows the local correlation for the STD simulation. Fixed STD window sizes have been simulated, with $\overline{n}_{s,w}$ values ranging from 1 to 20, indicated by the "$n =$" in the figure. It means, after simulating $\overline{n}_{s,w}$ values in an STD window, a new STD window is started. For the $n = 1$ plot this means, a pure static simulation is represented. This results in the correlation

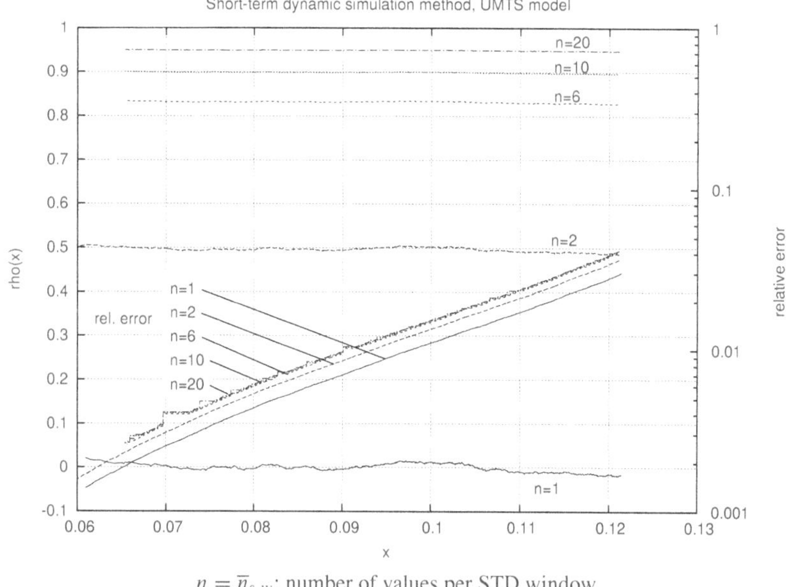


$n = \overline{n}_{s,w}$: number of values per STD window

**Figure 6.11:** Local correlation vs. UMTS load, short-term dynamic simulation

around 0. In the corresponding figure 5.14, the STD window sizes are given in time units, meaning that the *STD 1* plot does not directly correspond to the $n = 1$ plot here.

Large STD windows, leading to a smaller $p_{ex}$, result in higher local correlation. This agrees with the analytical calculations in figure 5.14. For $n = 20$, the correlation is again close to the correlation in figure 6.10. Considerations concerning the relative error of the simulation results and the impact on the simulation run-time are discussed below.

Next, figure 6.12 shows multi-step local correlation for the dynamic simulation. The plot for $n = 1$ shows the single-step local correlation, exactly the same as in figure 6.10. All other plots show the correlation which has been obtained by evaluating every second value (for $n = 2$), every sixth value (for $n = 6$), and so forth. This is also the reason for the higher variations of the curves for higher $n$, since the same simulation trace has been evaluated but with a smaller number of values. It could be considered to use the unused values for evaluation by doing a second pass over the sample starting over with the second value for $n = 2$, and even more passes for the higher $n$. This would, however, break the correlation at the wrap-around, inducing an additional unknown error. Compared to figure 5.18, the local correlation is higher for all step sizes, but the tendency is the same.

Next, figure 6.13 shows the group mean local correlation for the dynamic simulation. As introduced in sections 5.3.3.4 to 5.3.3.6, the mean value is calculated for each group of the given size

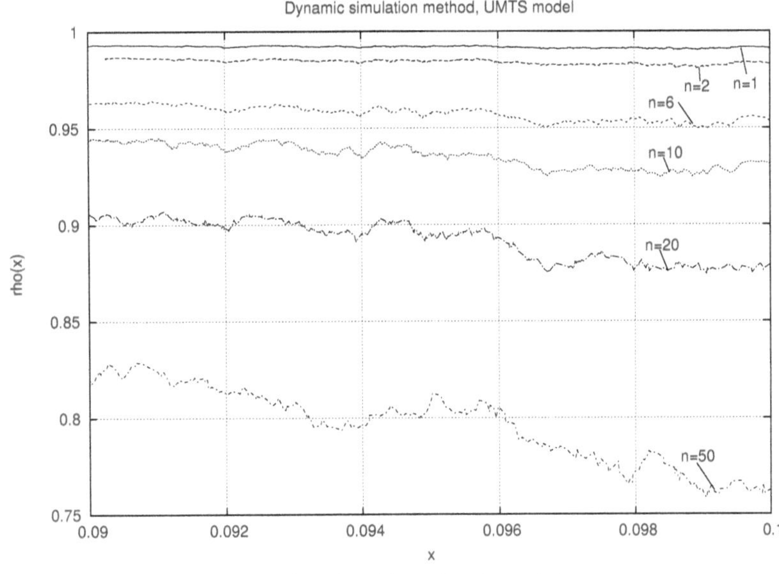


$n$: step-size of multi-step local correlation

**Figure 6.12:** Multi-step local correlation vs. UMTS load, dynamic simulation

$n$ in the sequence of groups. The correlation within this sequence of mean values is considered. Here, only the first 10,000 groups have been evaluated, which means, that all of the 200,000 values have been considered for the $n = 20$ plot, only the first half of them for the $n = 10$ plot, and so forth. This has been done to have the same number of group-mean values for the evaluation. Otherwise, the $n = 1$ plot would have been exactly the same as in figures 6.10 and 6.12, since for $n = 1$ it is again a single-step correlation. Compared to figure 5.21, again the tendency is the same, but the correlation here is much higher.

Finally, figure 6.14 shows the group mean local correlation for the short-term dynamic simulation. Obviously, as expected, the group-means are not correlated. The values are not all exactly 0, but they are very small. This results from an error limit which cannot be infinitely small. Further, the pseudo random number generators are not perfect in generating a totally uncorrelated random sequence, and thus, a small correlation will remain even between independent value traces. The curves for all $n$ are very close to each other because the same STD simulation trace was used for all different $n$. Always the first $n$ values of the STD windows have been evaluated.

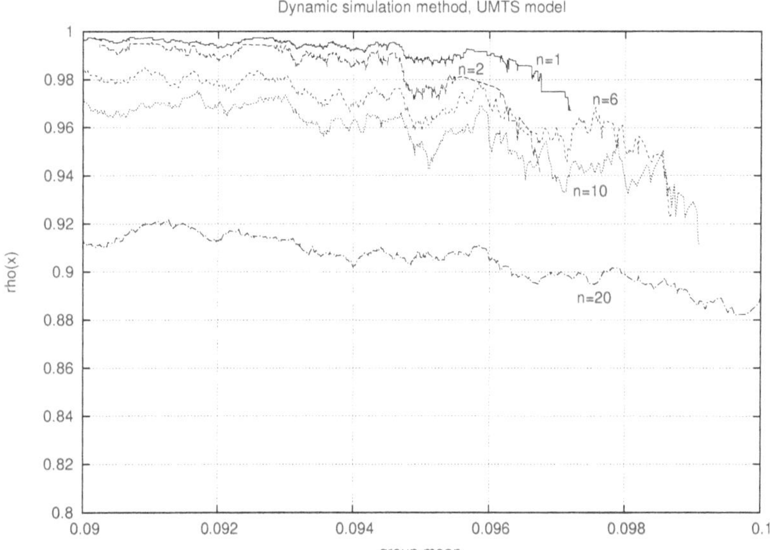


**Figure 6.13:** Group-mean local correlation (group size $n$), dynamic simulation

### 6.5.3 STD convergence

In section 6.5.1, the choice of an indicator for convergence of the evaluation has been explained. The scenario used is the one of section 6.1.3. In this section, simulations have been performed with the focus on the simulation time needed to reach convergence. According to the investigations in section 5.3, it is expected that the simulations with smaller STD windows converge more quickly to the given convergence criteria than the simulations with larger STD windows and the dynamic simulation. For the dynamic simulations, the same simulation toolkit is used with the only difference that there is one single STD window which lasts until the simulation ends.

#### 6.5.3.1 Voice model

Just like in section 6.5.2, the first investigations have been performed with a voice-only configuration with 500 users. Since the parameter for the on- and off-phases in up- and downlink are the same, only one link needs to be considered here, of which the uplink has been chosen. The cells of the central site (cells 0, 1 and 2) have been used as convergence indicators.

In figure 6.15, it is plotted the development of the mean value of the simulated sample up to the simulation time, meaning all values obtained from the start of the simulation are included in the

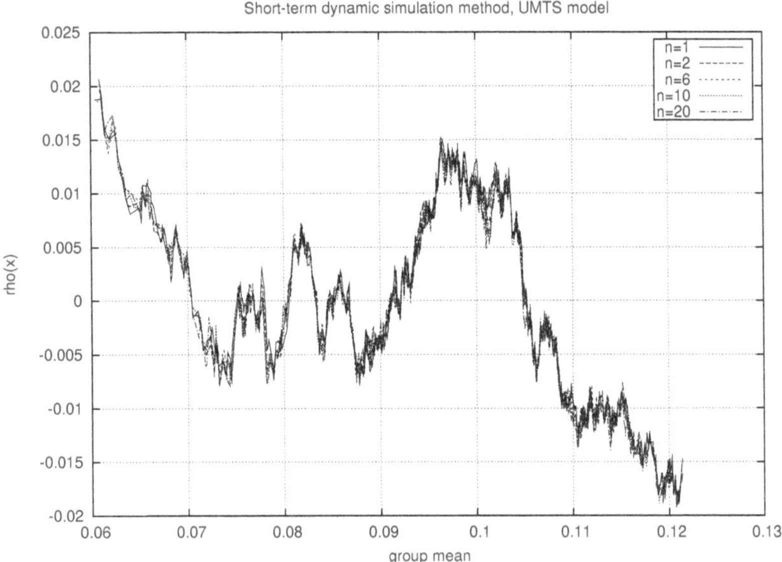


**Figure 6.14:** Group-mean local correlation (group size $n$), STD simulation

calculation of the mean value. The $x$-axis shows the simulation time in logarithmic scale, and the relative deviation in [%] from the mean value of the uplink load is shown on the $y$-axis.

On the left, figure 6.15 a) shows the development for the different STD window sizes. As expected, the speed with which the curves approach the zero deviation line increases from dynamic simulation over the simulation with large STD windows (100 s) to the simulation with smaller ones (10 s). On the right, figure 6.15 b) confirms that the slow convergence of the dynamic simulation not only is a special case for cell 0, but it applies also for the other two cells.

#### 6.5.3.2 Convergence evaluation

Different from the simulations in chapter 5, the LRE has not been used to evaluate the convergence. The slow convergence and the large deviation from the actual mean value can lead in some cases with large STD windows and especially in the dynamic simulation to a misinterpreted early convergence. An example is the dynamic simulation with the indicator cell 0 in figure 6.15 where the convergence has been identified at the simulation time of 952 s with the mean value being close to its lowest value. A detailed description of this example is given in appendix B.

In that example, a relative error limit of 10 % has been reached far too early (at 952 s) with its simulated mean value more than 15% lower than the actual mean value. For an error limit of

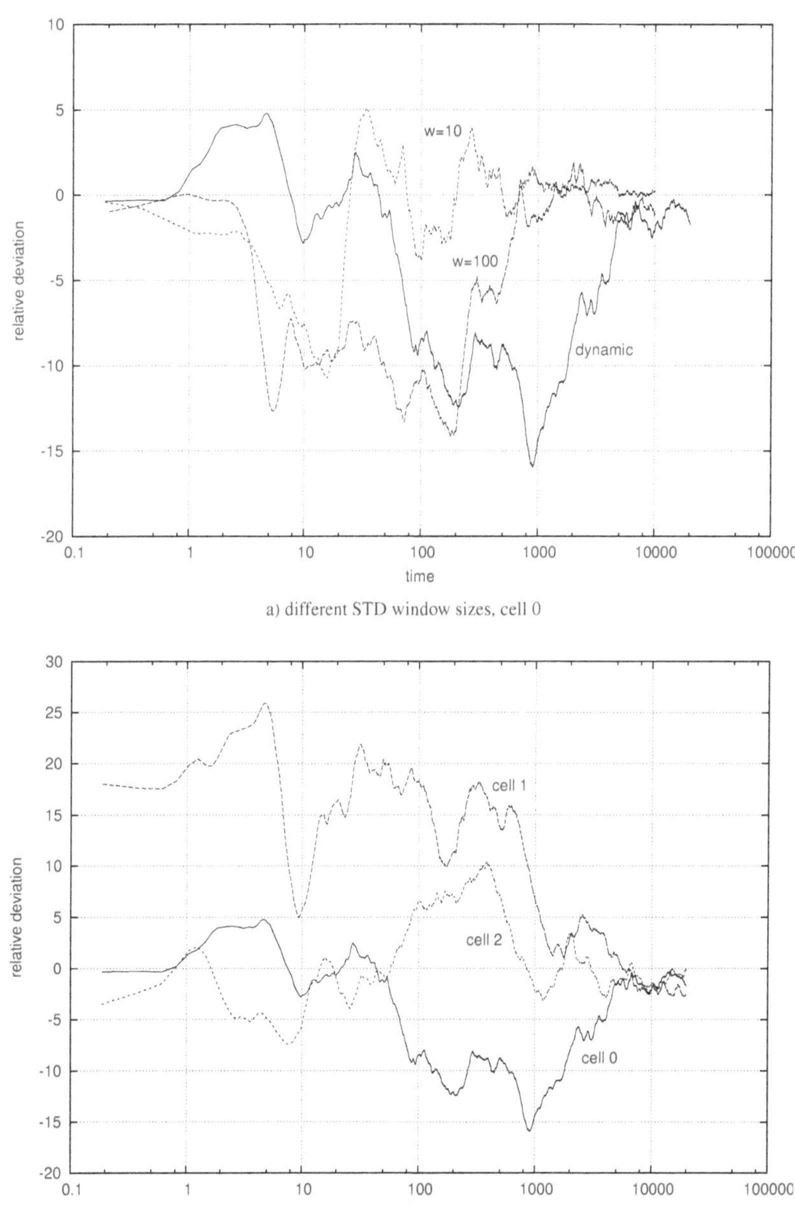


a) different STD window sizes, cell 0



b) dynamic simulation, different cells

**Figure 6.15:** Mean value development

5 %, the deviation has reduced to 8 %, and with a limit of 2 %, the deviation was far below 1 %. In contrast to this, all STD simulations even with the quite large STD window size of 100, for all mentioned error limits the deviation from the actual mean value never has been far above 1 %. Obviously, to achieve accurate results, especially for the dynamic simulation, the specification of the error limit can be a sensitive task, provided the actual mean value is a-priori unknown. It is not desirable to choose a very low error limit only to avoid large deviations from correct values due to non-representative simulated sequences. In the example, the fast but only virtual convergence gives the wrong impression of a fast simulation with sufficiently accurate results.

To avoid such misinterpretations in the convergence evaluation in this section, another mechanism has been used to which the actual mean value of the convergence indicator is a-priori known. This actual mean value has been obtained by a long static simulation with very low error limit.

As the convergence point, it has been defined the time at which the simulated mean value has been for a certain time uninterruptedly within a certain interval $\Delta_m$ around the (a-priori known) actual mean value. The chosen interval is here $\Delta_m = 1\,\%$ which means the simulated mean value has to stay between 99 % and 101 % of the actual mean value. The time it has to stay there uninterruptedly has been chosen to be relative to the elapsed simulation time at which the simulated mean has entered the interval the last time, or relative to the end of the initial phase, either of which comes later. It is called the relative duration $t_{rel}$ and is specified in [%]. During the initial phase $t_i$, the mean value is collected but not checked against the interval condition, which is to reach a certain minimum sample size including the statistical significance before evaluating the sample.

E. g., if the mean value enters the interval at 653 s and the required relative duration $t_{rel} = 10\,\%$, the convergence point is reached at 718 s if the mean value does not leave the interval up to that time. If the simulated mean stays within the interval already from 10 s while the initial phase $t_i = 200$ s, the condition with the 10 % will be reached at 220 s.

The method depending on the a-priori knowledge of the indicator mean value cannot be used in exactly the same manner for run-time control purposes as the application of the LRE. Without knowing anything about the statistical properties of the evaluated random sequence, the method cannot decide whether the simulation has provided satisfactory results and can therefore be terminated. The mean value of the convergence indicator, however, can be determined by a static simulation and with using the LRE. It should be avoided to spend too much time with this pre-simulation, but the interval chosen for the convergence evaluation should be much larger than the relative error reached with the static pre-simulation. Otherwise a simulation of which the simulated mean value approaches the actual mean will be most of the time outside the interval around the pre-simulated mean value of the static simulation.

#### 6.5.3.3 Convergence results

Given in tables 6.3 and 6.4 are the simulation time until convergence. Some figures could only be estimated since the convergence had not been reached within the simulation which has produced a long but finite random sequence. Thus, the bold figures represent numbers above the given value.

As expected, the results show the tendency to reach the convergence points later for increasing STD window size including the dynamic simulation considered as the largest. In the results of table 6.3, the initial phase $t_i$ has been set to $t_i = 100\,s$ (left part) and $t_i = 400\,s$ (right part). In both cases, the interval $\Delta_m$ has been set to 5 %, 2 %, and 1 %. Overall, the stability of the tendency is higher for the longer initial phase, and it increases with the smaller intervals. The stability of this tendency is small for small relative durations of $t_{rel} = 10\,\%$ and $t_{rel} = 20\,\%$, and it increases for $t_{rel} \geq 50\,\%$.

In table 6.4, only the highest accuracy ($\Delta_m = 1\,\%$) is shown. The right part with an initial phase of $t_i = 400\,s$ repeats the corresponding part of table 6.3, in the left part, $t_i = 200\,s$ has been applied. Compared to the 400 s case, only the simulations with the small STD windows and the small $t_{rel}$ are affected.

To have a fair comparison for all STD window sizes, $t_i$ must be the same for all. For very small STD windows, however, the convergence is faster, and the table gives the impression of saturation for $t_w \leq 1$ for the interval $\Delta_m = 1\,\%$, even more for the larger intervals. To compare convergence for small STD windows, the $t_i$ needs to be smaller.

### 6.5.4 STD window size

As has been found out, significant speed-up versus dynamic simulation can be reached with small STD windows. This has been proven for a theoretical model in section 5.3, and also for the simulation in this section, this statement is valid, apart from some limitations due to the complexity of the simulated system.

Depending on the kind of evaluation the simulation is focused on, see section 6.3, the STD window size may have to be quite large. An example is to evaluate the total session delay of a certain PS service, and only full sessions are of interest which started after the initial snapshot and finished before the end of the STD window. Furthermore, the variation of the amount of data to be transmitted is high, resulting in sub-exponentially decaying distributions of the session duration, especially if additionally the traffic load is high. For this situation, quite large STD windows are needed.

It can be easily calculated how large the STD windows need to be at minimum if the distribution function of the time based parameter is known and also the limit up to which a full session must be able to be evaluated. If, e. g., the voice service with a negative exponentially distributed session duration with a mean value of 120 s is evaluated and the simulation user wants 90 % of the calls to be included, the minimum STD window size must be 276 s. With an STD window

(a) 5% interval

| $t_w$ \ $t_{rel}$ | $t_i = 100$ s 10% | 20% | 50% | 100% | $t_i = 400$ s 10% | 20% | 50% | 100% |
|---|---|---|---|---|---|---|---|---|
| 0.1 s | 110 | 120 | 150 | 200 | 440 | 480 | 600 | 800 |
| 1 s | 110 | 120 | 150 | 200 | 440 | 480 | 600 | 800 |
| 10 s | 110 | 120 | 150 | 200 | 440 | 480 | 600 | 800 |
| 100 s | 152 | 476 | 614 | 819 | 440 | 480 | 614 | 819 |
| 1000 s | 130 | 142 | 177 | 261 | 440 | 480 | 1107 | 1476 |
| dynamic | 646 | 705 | 881 | 1174 | 646 | 705 | 881 | 1174 |

(b) 2% interval

| $t_w$ \ $t_{rel}$ | $t_i = 100$ s 10% | 20% | 50% | 100% | $t_i = 400$ s 10% | 20% | 50% | 100% |
|---|---|---|---|---|---|---|---|---|
| 0.1 s | 110 | 120 | 150 | 200 | 440 | 480 | 600 | 800 |
| 1 s | 110 | 120 | 150 | 200 | 440 | 480 | 600 | 800 |
| 10 s | 115 | 126 | 157 | 209 | 440 | 480 | 600 | 800 |
| 100 s | 682 | 744 | 930 | 1240 | 682 | 744 | 930 | 1240 |
| 1000 s | 194 | 212 | 1355 | 2735 | 994 | 1084 | 1355 | 2735 |
| dynamic | 763 | 833 | 1041 | 8488 | 763 | 833 | 1041 | 8488 |

(c) 1% interval

| $t_w$ \ $t_{rel}$ | $t_i = 100$ s 10% | 20% | 50% | 100% | $t_i = 400$ s 10% | 20% | 50% | 100% |
|---|---|---|---|---|---|---|---|---|
| 0.1 s | 110 | 120 | 150 | 200 | 440 | 480 | 600 | 800 |
| 1 s | 130 | 144 | 180 | 240 | 440 | 480 | 600 | 800 |
| 10 s | 117 | 162 | 947 | 1263 | 509 | 739 | 947 | 1263 |
| 100 s | 718 | 1317 | 1706 | 8874 | 718 | 1317 | 1706 | 8874 |
| 1000 s | 205 | 224 | 2721 | **15000** | 1086 | 1698 | 2721 | **15000** |
| dynamic | 904 | 1769 | 18356 | **25000** | 904 | 6024 | 18356 | **25000** |

Long-term mean values for cells 0, 1 and 2:
$\rho_0 = 0.09060821345$, $\rho_1 = 0.09057746281$, $\rho_2 = 0.09016453876$,
bold figures represent a number above the given value,
$t_w$: STD window size, $t_{rel}$: relative duration, $t_i$: initial phase

**Table 6.3:** Convergence points in [s], different $t_i$, different intervals $\Delta_m$

| $t_{rel}$ / $t_w$ | $t_i = 200\,s$ | | | | $t_i = 400\,s$ | | | |
|---|---|---|---|---|---|---|---|---|
| | 10% | 20% | 50% | 100% | 10% | 20% | 50% | 100% |
| 0.1 s | 220 | 240 | 300 | 400 | 440 | 480 | 600 | 800 |
| 1 s | 220 | 240 | 300 | 400 | 440 | 480 | 600 | 800 |
| 10 s | 253 | 739 | 947 | 1263 | 509 | 739 | 947 | 1263 |
| 100 s | 718 | 1317 | 1706 | 8874 | 718 | 1317 | 1706 | 8874 |
| 1000 s | 220 | 1698 | 2721 | **15000** | 1086 | 1698 | 2721 | **15000** |
| dynamic | 904 | 6024 | 18356 | **25000** | 904 | 6024 | 18356 | **25000** |

**Table 6.4:** Convergence points in [s], different $t_i$, interval $\Delta_m = 1\,\%$

size of only 120 s, only 63 % of the possible calls will be included. This applies only for the sessions starting immediately after the initial snapshot, while sessions starting later will be included with a smaller probability. Another effect has to be paid attention to: Since unfinished sessions at the end of the STD window are not included into the evaluation, the session duration of the evaluated sessions is not representative because long sessions are disregarded. This has to be taken into account if information about longer sessions is important, at least in an interpolated manner.

Many other possible statistics can be evaluated with small STD windows. E. g., for the dwell time in SHO regions (SHO duration), if not evaluated session based, the STD windows can be in the order of seconds if the users are moving fast and the SHO regions are not too large.

Dropping can be evaluated with practically any STD window size. Any session being active in the initial snapshot can be dropped in one of the very next moments if some of the dynamic changes have a sufficient impact on the load situation of the corresponding user, e. g., if a movement into a more critical region has been performed.

In section 6.3.4, examination of detailed RRM is discussed. One way to evaluate such details is to trace several types of events and parameters. Information like the effort needed to resolve a certain traffic load situation, the frequency of situations which require such effort, and many others, can be obtained with simulations with small STD windows.

## 6.6 Speed-up by event grouping

In an STD simulation, the system recalculation principally has to be performed after every event, according to figure 6.3 and the *process input* action of figure 5.1. This is very frequently, but it is required to keep the system in a consistent state. Otherwise it would be possible for a user to move into an area where no radio coverage exists or where the user would need to be connected to another cell, etc. The system recalculation, however, consumes the major part of the simulation run-time of an STD simulation.

Since, in contrast to static simulation in which the difference between two consecutive system states is usually high because of the independence, in STD simulations only very small changes are aligned with a single event. The impact of such changes on the system parameters which in turn have impact on the performance of other users or the network, can be small, but it depends on the type of event. The impact on the cell load, which is incorporated into all RRC decisions, is quite small if only a voice user changes the activity. The impact is higher if a user enters resp. leaves the scenario, or if a user moves.

The amount of changes the event causes to the system, or the relevance of the event, depends on further parameters like the service type in connection with the radio bearer and on the pixel size which determines the travelled distance for every mobility event.

Especially for larger scenarios, a system recalculation takes a long time while a single event causes only small impact to the system which is also limited to local areas. The distance up to which the impact of a single event reaches cannot be easily predicted, but in large scenarios not all cells will be influenced.

### 6.6.1 Grouping and weighting

An idea is now to collect some of the changes caused by the events and perform a system recalculation not after every single event. A simple way to do so is, to have a fixed number $T$, the so-called threshold, up to which the events are counted (grouped) before a system recalculation is performed.

A more sophisticated approach is to assign a weight $w(\mathbf{e}_i)$ to an event with $\mathbf{e}_i$ being an event with the index $i$, and to accumulate these weights up to a threshold $T$, equation (6.7):

$$\sum_{i=c_l+1}^{c_c} w(\mathbf{e}_i) \geq T \tag{6.7}$$

Here, $c_l$ resp. $c_c$ mean the index of the event after which the last recalculation has taken place resp. the index of the current (latest) event. After a recalculation, the index value $c_l$ is updated to $c_c$. In the simple approach only counting the events, it is $w(\mathbf{e}_i) = w = 1$.

Starting with the distinction of two different classes of events, the *heavy* events representing significant changes, and the *light* events. A light event is assigned a weight of $w(\mathbf{e}_L) = 1$, and a heavy event has a weight of $w(\mathbf{e}_H) = T$. This makes a heavy event initiate a system recalculation immediately, since every time a heavy event occurs, the threshold $T$ is reached.

With this setup, it is possible to predict the total number of system recalculations $N_R$ that will be performed in a simulation with a total number of $N$ events and a threshold $T$. The probability $p_H$ of a heavy event can be assumed as independent of the class of the previous events, resulting in a geometrically distributed number of light events until the next heavy event occurs. If $p_H$ is high, the gain by saving on system recalculation will be small. Equation (6.8) shows the derivation for $N_R$:

$$\begin{aligned}
N_R &= N_H + N_H \cdot \mathrm{E}[X_R] \\
&= N_H \cdot (1 + \mathrm{E}[X_R]) \\
\mathrm{E}[X_R] &= \sum_{x=1}^{\infty} \left\lfloor \frac{x}{T} \right\rfloor P(X_H = x) \\
P(X_H = x) &= p_H \cdot p_L^x \\
p_L &= 1 - p_H \\
p_H &= \frac{N_H}{N_H + N_L} = \frac{N_H}{N} \\
N &= N_H + N_L
\end{aligned} \tag{6.8}$$

This finally leads to equation (6.9):

$$N_R = N_H \cdot \left(1 + \sum_{x=1}^{\infty} \left\lfloor \frac{x}{T} \right\rfloor \cdot \frac{N_H}{N} \cdot \left(1 - \frac{N_H}{N}\right)^x\right) \tag{6.9}$$

$X_R$ means the RV for the number of recalculations between 2 heavy events, $X_H$ is the RV for the number of events from last heavy event to the next heavy event, and $N_H$ resp. $N_L$ are the total number of heavy resp. light events. $p_H$ resp. $p_L$ are the probabilities of a heavy resp. a light event.

### 6.6.2 Simulation results

The probability of a heavy event $p_H$ depends on the scenario and on the classification. For a relatively simple scenario with only voice users, user arrivals, session terminations, and mobility events should be considered as heavy events, while activity changes should be the light events. This has been applied to the example simulation with the results shown in table 6.5.

| Type of event | Event count | |
|---|---|---|
| | 20 m | 5 m |
| Activity changes | 4235 | 4683 |
| Movements | 87 | 352 |
| New user arrivals | 29 | 25 |
| Call terminations | 31 | 33 |
| Light events $N_L$ | 4235 | 4683 |
| Heavy events $N_H$ | 147 | 410 |
| $p_H$ | 3.35 % | 8.05 % |

**Table 6.5:** Occurring number of events during two example simulations

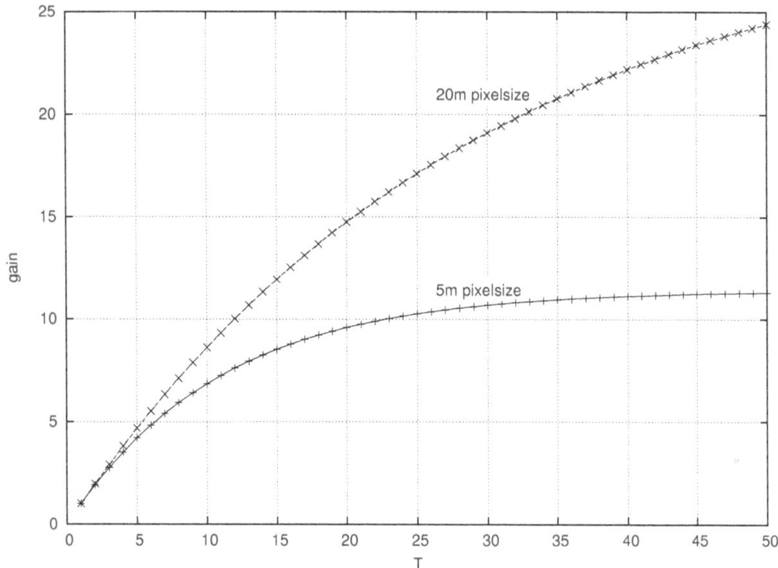


**Figure 6.16:** Achievable gain vs. threshold $T$ with event grouping

Pixel sizes of 5 m and 20 m have been used. It has been taken care of a comparable total travelled distance of the users to demonstrate the influence of the higher number of mobility events for smaller pixel sizes. Figure 6.16 shows the gain that is achieved by applying event grouping for the two examples of table 6.5 for different thresholds $T$. As with non-grouped simulations $N_R = N$, the gain is defined to be as in equation (6.10):

$$gain = \frac{N}{N_R} \tag{6.10}$$

For the plot, the formula of equation (6.9) has been applied on the basis of the $p_H$ derived from the example simulations.

Larger $T$ will not lead to further significant increase of the gain. This is because the probability to have at least $T-1$ successive light events before a heavy event occurs decreases with increasing $T$. This effect is stronger for the smaller pixel size because of the larger number of mobility events. It leads to a higher percentage of heavy events $p_H$. For the figure, the formula has been applied with the empirical $p_H$ from the simulation, and the deviation from the actually acquired number of system recalculations in the corresponding simulations is below 0.3 % for threshold values up to $T = 10$.

Concluding, the formula can be used to estimate the gain which a grouping with a certain threshold $T$ could achieve with a $p_H$ that has been obtained by short simulations of a typical configuration of the investigated scenarios. In case of an expected gain, the application of event grouping can save significant amounts of time. Longer simulations also with different scenarios have shown the accuracy of the results do not suffer from the application of event grouping up to a threshold of $T = 10$.

It is relevant here to consider the accuracy from the perspective of the network behaviour. In the evaluation of QoS parameters, the decisions of the RRM are most important, leading to blocking, dropping, soft-handover, etc. On these decisions, however, the delayed inclusion of the *light* events can only have significant influence in situations of very high load where very small load changes can lead to crossing of load limits.

#### 6.6.2.1 Outlook

Even more sophisticated would be to have a finer classification of the events. The weight of the activity changes of PS calls would depend on the radio bearer and its estimated load contribution, and the PS activity changes would further be classified rather like user arrivals and call terminations because new resources are occupied resp. occupied resources are freed in contrast to CS sessions. Bearer dependence should also be taken into account for CS sessions, if applicable. It is also important to classify the mobility events according to the pixel size. Smaller pixels lead to a larger number of mobility events, but each of these have smaller impact due to the smaller distance travelled for a single pixel. The velocity of a user could also be considered since the requirements concerning the signal quality are higher for users travelling with higher speed, resulting in a higher load contribution for such users. This applies also for a distinction between the service types which have, in general, different requirements concerning the signal quality, independent of the speed. Since the system relevance of the events would be considered more precisely, these improvements would eliminate some of the uncertainty discussed in the previous paragraph.

## 6.7 Conclusions

It has been shown how the STD simulation concept can be applied to the domain of network planning for $3^{rd}$ generation mobile networks such as UMTS. Possible targets for evaluation have been identified, and it has been explained why it is reasonable to use the STD simulation and what has to be taken into account.

In the investigation of the performance of the STD simulation concept, it has been shown that not only for simple models as in section 5.3 a significant speep-up can be achieved with small STD windows, but also for simulations of realistic UMTS scenarios.

It has been identified that for some evaluation targets the STD simulation makes sense even for very small STD windows with accordingly high gain.

# Chapter 7

# Conclusions

In this work, the basic idea of simulation speed-up is highlighted. The term *statistical accuracy* is introduced to allow a comparison of reference simulation techniques with speed-up approaches. One important factor for statistical accuracy is the state space coverage and its sources. Several possibilities to increase the statistical accuracy by changing parameters influencing the coverage are identified. The evaluation method LRE is described as the means for evaluation and simulation control for the simulation methods addressed in this work.

Different simulation speed-up approaches are examined. These are RESTART as a method to speed-up simulations of rare events, general parallelisation, and the new short-term dynamic simulation concept. Combinations of these approaches are investigated to different extents.

The basic principle of the rare event simulation technique RESTART is introduced. The importance function is considered with respect to its usage in the RESTART simulation toolkit MuSICS. It is shown how multivariate importance functions can be applied together with the LRE as the evaluation and simulation control mechanism, and further, how the functionality of the importance function can be separated from the evaluation of the target random variable.

It is discussed how to handle transition states in single-step mode. Usually, the linear selection should be applied, and a state limitation can be applied if incorrect configurations cannot be eliminated completely. Predictions of the needed number of transition states are based on assumptions which in general can be not justifiable. It is recommended to perform pre-simulations if such information is needed for optimisation purposes. Pre-simulations with single-step combined with subsequent global-step simulation saves time and memory and this is therefore recommended as a combined strategy.

One of the shortcomings of the RESTART mechanism is the sometimes limited placement possibility of the thresholds. The ability of refining the thresholds allows placing them closer to the optimum in cases with regions of steep decline in the complementary cumulative distribution function, especially at the origin of the evaluated value space. Configured carefully it provides remarkable speed-up together with adequate accuracy.

One of the main contributions of this work is the in-depth investigation of the new short-term dynamic (STD) simulation concept. After describing the short-term dynamic simulation concept in detail, an analytical proof on the basis of a simple queueing model shows that the concept is able to speed-up simulations considerably. This is achieved by a reduction of the local correlation without changing the behaviour of the system model. The best speed-up is, however, achieved with small STD windows, while large STD windows reflect a behaviour with a tendency towards dynamic simulations.

For the application of the short-term dynamic simulation concept to the network planning in UMTS, the concept of a simulation toolkit is described. A method to reduce the number of system recalculations has been found to reduce the simulation run-time in many cases without reducing the quality of the results.

The performance of the concept is examined on the basis of UMTS models, which are more complex than the simple model used for the analytical investigation. The results of the analytical evaluation and the simple queueing network are confirmed. The evaluation of the simulation methods and scenarios further confirm a gain as identified in the analytical investigation also for these complex models.

It is shown that certain combinations of the different speed-up methods addressed in this work have a remarkable speed-up potential. With a fast network interconnection technology, the combination of RESTART and parallelisation provides a performance gain also on networks of workstations. Because of the communication overhead, however, on a network of workstations, the combination requires fast hardware solutions. On SMP systems with more than two processors per node, better performance results are achieved. Resulting from the relatively small optimum splitting level of the RESTART method, the scalability of this combination is limited.

Combining the STD simulation concept with parallelisation is principally intuitive. The merging of the results of the STD windows, however, has to be taken into account. For large STD windows, the scalability is limited, but the effort for communicating the results is low in contrast to cases with small STD windows. For those, the scalability is better.

A critical issue when combining the STD simulation concept with RESTART is the lack of information about past events. This history information is incorporated in the transition states of RESTART, but the initial snapshots of the STD windows are generated without such data. On the other hand, the STD simulation can create snapshots as needed according to statistical properties, while in RESTART such system states have to be obtained during the simulation. This enables the STD simulation to force a simulation in a defined target importance region. Without a history of the initial snapshots, however, many of them will not contribute to the target evaluation. Defining a hysteresis region can help to compensate this effect.

This work has shown that the STD simulation concept is well suited to increase the statistical accuracy. The method can be applied if representative independent snapshots can be generated with the same statistical properties as the STD window. This requirement is proven to be fulfilled for simple models as it can be assumed for mobile network models. Examples from the area of UMTS network planning have been presented where the STD method allowed the evaluation of dynamic performance measures profiting from the speed-up of this method.

# Appendix A

# Mathematical Derivations

## Extra loop transition at state 0 in DTMC

In section 5.3.1.1, it is explained why a DTMC model has been chosen which has an extra loop transition at state 0. The transition probability has been named $p_{s0}$. Now, it is shown the derivation of equation (5.1). For clearness, figure 5.5 has been repeated here as figure A.1.

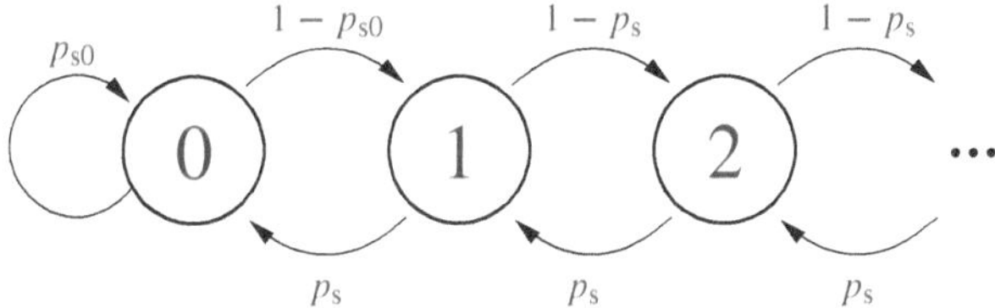


**Figure A.1:** DTMC of M/M/1 model with extra loop transition at state 0

Equations (A.1) to (A.4) are according to equations (5.6) to (5.8) where equation (5.1) has already been incorporated.

$$P(0) = P(0) \cdot p_{s0} + P(1) \cdot p_s \tag{A.1}$$

$$P(1) = P(0) \cdot (1 - p_{s0}) + P(2) \cdot p_s \tag{A.2}$$

$$P(x) = P(x-1) \cdot (1 - p_s) + P(x+1) \cdot p_s \quad \text{for } x = 2 \ldots \infty \tag{A.3}$$

$$\sum_x P(x) = 1 \tag{A.4}$$

Equations (A.1) and (A.2) lead to

$$\begin{aligned} P(0) &= P(1) \cdot \frac{p_s}{1 - p_{s0}} && \text{(A.5)} \\ P(1) &= P(2) \cdot \frac{p_s}{1 - p_s} && \text{(A.6)} \end{aligned}$$

By induction, according to equation (5.9), equations (A.5) and (A.6) lead to

$$P(x) = P(x-1) \cdot \begin{cases} \frac{1-p_{s0}}{p_s} & \text{for } x = 1 \\ \frac{1-p_s}{p_s} & \text{for } x \geq 2 \end{cases} \tag{A.7}$$

This leads to

$$P(x) = \left(\frac{1-p_s}{p_s}\right)^{x-1} \cdot \frac{1-p_{s0}}{p_s} \cdot P(0) \qquad \text{for } x \geq 1 \tag{A.8}$$

Together with equation (A.4), $P(0)$ can be calculated as follows

$$\begin{aligned} & P(0) + \sum_{x=1}^{\infty} \left(\frac{1-p_s}{p_s}\right)^{x-1} \cdot \frac{1-p_{s0}}{p_s} \cdot P(0) = 1 \\ \Rightarrow \quad & P(0) \cdot \left(1 + \frac{1-p_{s0}}{p_s} \cdot \sum_{x=1}^{\infty} \left(\frac{1-p_s}{p_s}\right)^x\right) = 1 \\ \Rightarrow \quad & P(0) \cdot \left(1 + \frac{1-p_{s0}}{p_s} \cdot \frac{1}{1 - \frac{1-p_s}{p_s}}\right) = 1 \\ \Rightarrow \quad & P(0) \cdot \frac{2p_s - p_{s0}}{2p_s - 1} = 1 \\ \Rightarrow \quad & P(0) = \frac{2p_s - 1}{2p_s - p_{s0}} && \text{(A.9)} \end{aligned}$$

Together with equation (A.8) and according to equation (5.10), this leads to

$$P(x) = \frac{2p_s - 1}{2p_s - p_{s0}} \cdot \frac{1-p_{s0}}{p_s} \cdot \left(\frac{1-p_s}{p_s}\right)^{x-1} \tag{A.10}$$

In the comparison model, the M/M/1 model considered as CTMC as in figure 5.6, the state probabilities are as follows

$$\begin{aligned} P_{\text{ref}}(0) &= 1 - \eta && \text{(A.11)} \\ P_{\text{ref}}(x) &= (1 - \eta) \cdot \eta^x \end{aligned}$$

According to equation (5.5), $p_s = \frac{1}{1+\eta}$. With this, equation (A.9) becomes

$$P(0) = \frac{\frac{2}{1+\eta} - 1}{\frac{2}{1+\eta} - p_{s0}} = \frac{1 - \eta}{2 - p_{s0} \cdot (1 + \eta)}. \tag{A.12}$$

To fulfil $P_{ref}(0) = P(0)$, it has to be

$$\begin{aligned} & 2 - p_{s0} \cdot (1 + \eta) \stackrel{!}{=} 1 \\ \Rightarrow \quad & p_{s0} = \frac{1}{1 + \eta} = p_s \quad \square \end{aligned} \tag{A.13}$$

# Appendix B

# LRE example behaviour

The phenomenon leading to the fast convergence of the dynamic simulation in table can be explained as follows: The point of simulation time at which the detection of convergence has taken place was at 952 s, exactly the time when in the figure the corresponding plot turns to raise again from its absolute minimum.

Since the mean value development is considered, the divergent behaviour of the curve can be explained with a value sequence with values far below the static mean value of 0.0906082 especially for the majority of the simulation time up to point of 952 s.

In fact, the evaluation process has received a sample of a certain size with a mean value far below the steady state mean value of the model. This mean value is, however, not known a-priori to the evaluation process. Thus, if the aspects building the error calculation lead to an error below the given maximum, the evaluation process cannot be accused of misinterpretation of the given simulation sample.

The fact that the convergence has been reached a short time after the simulated mean value has reached its absolute minimum, can be explained with the large sample conditions, see section 2.4 and especially equation (2.4). The first large sample condition in equation (2.2) is fulfilled after 1000 observations have been made, which happens in the example after a little more than 2 s simulation time.

The second large sample condition in equation (2.3) is about the cumulative frequencies of the intervals and, thus, related to the aimed distribution function. Into these intervals the value space corresponding to the distribution function is divided. This is done in evaluation processes as the LRE either automatically or with given interval boundaries. The left part of the equation is about the $r_i$ and related to the distribution function $F(x)$ while the right part about $v_i$ is related to the complementary cumulative distribution function $G(x)$, and $i$ is the index of the interval. Both of them have to be larger than 100 for each $i$ which is critical only for the boundary intervals. This condition can be easily fulfilled if the probabilities of the boundary intervals are not too small which can be decided by setting the limit probability for the intervals to, e. g., 10%.

The last large sample condition in equation (2.4) is about the correlation. First, according to the left two parts of the equation, the transition frequencies have to be larger than 10 for both directions at every interval boundary. This assures that there is at least a minimum negative contribution to the local correlation at the interval boundary to prevent a local correlation of $+1.0$. Further, according to the right part of the equation, the difference between the cumulative frequency of the one side and the transition frequency to that side must be larger than 10. This is giving a lower bound contribution to the positive local correlation to prevent a local correlation of $-1.0$ at that boundary. This would mean that a transition to state $S_i$ from a state left resp. right of $S_i$ would be followed every time and immediately by a transition back to a state left resp. right of $S_i$.

In this special case, a jump from low values to higher values has been made which made the mean value increase again. After this jump, the missing number of transitions from the lowest values to the higher ones has been collected fulfilling equation (2.4) as the last criteria.

# Bibliography


[Bay70] A. J. Bayes. *Statistical Techniques for Simulation Models.* The Australian Computer Journal, Vol. 2(4), pp. 180–184, 1970.

[BFS87] Paul Bratley, Bennett L. Fox, Linus E. Schrage. *A Guide to Simulation.* Springer, second edition, 1987.

[ETS98] *Selection procedures for the choice of radio transmission technologies of the UMTS.* Technical Report, ETSI, April 1998.

[Fuj00] Richard M. Fujimoto. *Parallel and Distributed Simulation Systems*, chapter 7. John Wiley & Sons, 2000.

[Gar00] Marnix J.J. Garvels. *The splitting method in rare event simulation.* Phd Thesis, Universiteit Twente, 2000.

[GF98] C. Görg, O. Fuß. *Comparison and Optimization of RESTART Run Time Strategies.* AEÜ, Vol. 52, pp. 197–204, 1998.

[GLA99] C. Görg, E. Lamers, R.G. Addie. *Broadband Traffic Modeling: Rare Event Simulation for Gaussian and M/Pareto Processes with the same Autocovariance.* Workshop 11.-12. March 1999, University of Twente, Enschede, The Netherlands, 1999.

[GLS99] William Gropp, Ewing Lusk, Anthony Skjellum. *Using MPI – Portable Parallel Programming with the Message-Passig Interface.* The MIT Press, second edition, 1999.

[GS90] C. Görg, F. Schreiber. *Der LRE-Algorithmus für die statistische Auswertung korrelierter Zufallsdaten.* 6. ASIM-Symposium, Wien 1990; Breitenecker, F. et al., Simulationstechnik, Vieweg, Wiesbaden, pp. 170–174, 1990.

[GS96] C. Görg, F. Schreiber. *The RESTART/LRE Method for Rare Event Simulation.* In *1996 Winter Simulation Conference*, pp. 390–397, Coronado, California, USA, December 1996.

[Gör97] C. Görg. *Verkehrstheoretische Modelle und Stochastische Simulationstechniken zur Leistungsanalyse von Kommunikationsnetzen*, volume ABMT 13. Verlag der Augustinusbuchhandlung, Aachener Beiträge zur Mobil- und

Telekommunikation, Aachen, 1997. Habilitationsschrift, Lehrstuhl Kommunikationsnetze, RWTH Aachen.

[Hee95] P. E. Heegaard. *Comparison of speed-up techniques for simulation.* In *The 12th Nordic Teletraffic Seminar (NTS-12), August, 1995*, pp. 407–420, Trondheim, Norway, 1995.

[HT00] H. Holma, A. Toskala. *WCDMA for UMTS.* Wiley, 2000.

[JA93] M. Junius, S. Arefzadeh. *CNCL – ComNets Class Library.* RWTH Aachen, Lehrstuhl für Kommunikationsnetze, 1993.

[KWD99] Frederick Kuhl, Richard Weatherly, Judith Dahmann. *Creating Computer Simulation Systems, An Introduction to the High Level Architecture.* Prentice Hall, 1999.

[LG00] E. Lamers, C. Görg. *RESTART/LRE efficiency and feasibility for models with many aggregated sources.* Workshop 05.-06. October 2000, University of Pisa, Italy, 2000.

[LG02] E. Lamers, C. Görg. *Distributed RESTART with MuSICS: Rare Event Simulation on a Network of Workstations.* Workshop on Rare Event Simulation, Madrid, Spain, April 2002.

[LG04] E. Lamers, C. Görg. *RESTART: Distributed Simulations and Parameter Optimization.* In *Workshop on Rare Event Simulation*, Budapest University of Technology and Economics, Hungary, September 2004.

[LKG01] E. Lamers, A. Könsgen, C. Görg. *Investigations on the Coexistence of IEEE 802.11a and HiperLAN/2 by enhancing existent simulators and combining them via the HLA.* In *4th International Symposium on Wireless Personal Multimedia Communications*, Aalborg, Denmark, September 2001.

[LPTG03] E. Lamers, R. Perera, U. Türke, C. Görg. *Short Term Dynamic System Level Simulation Concepts for UMTS Network Planning.* In *Proc. 18th International Teletraffic Congress*, pp. 101–110, Berlin, Germany, Sep. 2003.

[Mom03] *http://momentum.zib.de*, 2003. IST-Project MOMENTUM (IST-2000-28088).

[PEF+02] R. Perera, A. Eisenblätter, E. R. Fledderus, C. Görg, M. Scheutzow, S. Verwijmeren. *Pixel Oriented Mobility Modelling for UMTS Network Simulations.* In *Proceeding of IST Mobile and Wireless Telecommunications Summit 2002*, pp. 828–831, Thessaloniki, Greece, June 2002.

[PLG02] R. Perera, E. Lamers, C. Görg. *Optimum Number of Retrials in Multi-Threshold RESTART Simulations.* Workshop on Rare Event Simulation, Madrid, Spain, April 2002.

[PLG+04] R. Perera, E. Lamers, C. Görg, U. Türke, T. Winter. *Analysis of Soft-Handover Characteristics in Realistic Urban UMTS Radio Networks.* In *Proc. of the IEEE Vehicular Technology Conference (VTC 2004-Spring), Milan, Italy*, May 2004.

[Rin97] Horst Rinne. *Taschenbuch der Statistik.* Verlag Harri Deutsch, 1997. 2. Auflage.

[Sch84] F. Schreiber. *Time efficient simulation: the LRE-algorithm for producing empirical distribution functions with limited relative error.* AEÜ, Vol. 38, pp. 93–98, 1984.

[Sch87] F. Schreiber. *The Empirical Stationary Distribution Function of Markovian Correlated Random Sequences.* AEÜ, Vol. 41, pp. 257–263, 1987.

[SG94] F. Schreiber, C. Görg. *Rare Event Simulation: a Modified RESTART-Method using the LRE-Algorithm.* Teletraffic and Datatraffic, Proceedings 14th ITC, Antibes, Juan-Les-Pins, France, June 6-10, 1994, pp. 787–796, 1994.

[SG96] F. Schreiber, C. Görg. *Stochastic Simulation: a Simplified LRE-Algorithm for Discrete Random Sequences.* AEÜ, Vol. 50, pp. 233–239, 1996.

[TPL+03] U. Türke, R. Perera, E. Lamers, T. Winter, C. Görg. *An Advanced Approach for QoS Analysis in UMTS Radio Network Planning.* In *Proc. of ITC 18*, Berlin, 2003.

[Tri02] Kishor S. Trivedi. *Probability and Statistics with Reliability, Queuing, and Computer Science Applications.* John Wiley & Sons, Inc., 2002.

[VA98] José Villén-Altamirano. *RESTART Method for the Case Where Rare Events Can Occur in Retrials From Any Threshold.* International Journal of Electronics and Communications, Vol. 52, No. 3, pp. 183–189, 1998.

[VAMMGFC94] M. Villén-Altamirano, A. Martínez-Marrón, J. Gamo, F. Fernández-Cuesta. *Enhancement of the Accelerated Simulation Method RESTART by considering Multiple Thresholds.* In *Proceedings 14th International Teletraffic Congress*, pp. 797–810. North-Holland, 1994.

[VAVA91a] M. Villén-Altamirano, J. Villén-Altamirano. *Accelerated simulation of rare events using RESTART method with hysteresis.* In J. Filipiak, editor, *Telecommunication Services for Developing Economies.*, pp. 675–686. Proceedings of the ITC Specialist Seminar, Cracow, Poland, April 1991, Elsevier, Amsterdam, Netherlands, 1991.

[VAVA91b] M. Villén-Altamirano, J. Villén-Altamirano. *RESTART: a Method for Accelerating Rare Event Simulations.* In J.W. Cohen, C.D. Pack, editors, *Queueing, Performance and Control in ATM*, pp. 71–76. 13th International Teletraffic Congress, Copenhagen, North-Holland, 1991.

[VAVA99] M. Villén-Altamirano, J. Villén-Altamirano. *About the Efficiency of RESTART*. Workshop 11.-12. March 1999, University of Twente, Enschede, The Netherlands, 1999.

[Wal01] Bernhard Walke. *Mobile Radio Networks - Networking and Protocols*. John Wiley & Sons, Chichester, New York, Wenheim, Brisbane, Singapore, Toronto, 892 pages, ISBN 0-471-49902-1, Nov 2001.

## Advanced Studies Mobile Research Center Bremen

Herausgeber: Prof. Dr. Otthein Herzog, Prof. Dr. Carmelita Görg, Prof. Dr.-Ing. Bernd Scholz-Reiter

Eugen Lamers

**Contributions to Simulation Speed-Up**

Rare Event Simulation and Short-Term Dynamic Simulation for Mobile Network Planning

2008. xxi, 148 pp. with 57 Fig. and 11 Tab, Dissertation Universität Bremen 2007

(Advanced Studies Mobile Research Center Bremen) Softc. EUR 45,90

ISBN 978-3-8348-0524-9

Ingrid Rügge

**Mobile Solutions**

Einsatzpotenziale, Nutzungsprobleme und Lösungsansätze

2008. XV, 319 S. mit 109 Abb. u. 8 Tab. Dissertation Universität Bremen, 2006

(Advanced Studies Mobile Research Center Bremen) Br. EUR 55,90

ISBN 978-3-8350-0919-6

Ulrich Türke

**Efficient Methods for WCDMA Radio Network Planning and Optimization**

2008. xxv, 165 pp. with 70 Fig. and 13 Tab. Dissertation Universität Bremen

(Advanced Studies Mobile Research Center Bremen) Softc. EUR 45,90

ISBN 978-3-8350-0903-5

Hendrik Witt

**User Interfaces for Wearable Computers**

Development and Evaluation

2008. xxi, 273 pp. with 101 Fig. and 7 Tab. Dissertation Universität Bremen, 2007

(Advanced Studies Mobile Research Center Bremen) Softc. EUR 49,90

ISBN 978-3-8351-0256-9

Abraham-Lincoln-Straße 46
65189 Wiesbaden
Fax 0611.7878-400
www.viewegteubner.de

Stand Mai 2008.
nderungen vorbehalten.
Erhältlich im Buchhandel oder im Verlag.

SPRINGER NATURE

# GPSR Compliance

*The European Union's (EU) General Product Safety Regulation (GPSR) is a set of rules that requires consumer products to be safe and our obligations to ensure this.*

*If you have any concerns about our products, you can contact us on ProductSafety@springernature.com*

In case Publisher is established outside the EU, the EU authorized representative is:

Springer Nature Customer Service Center GmbH
Europaplatz 3
69115 Heidelberg, Germany

Zeitfracht Medien GmbH
Ferdinand-Jühlke-Straße 7
99095 Erfurt, Deutschland
produktsicherheit@kolibri360.de